UPSCALING MULTIPHASE FLOW IN POROUS MEDIA

T0135379

Upscaling Multiphase Flow in Porous Media

From Pore to Core and Beyond

Edited by

D.B. DAS
University of Oxford, U.K.

and

S.M. HASSANIZADEH
Utrecht University, The Netherlands

*Part of this volume has been published in the Journal
Transport in Porous Media vol. 58, No. 1–2 (2005)*

 Springer

A C.I.P. Catalogue record for this book is available from the Library of Congress.

ISBN 978-90-481-6888-0 (PB)
e-ISBN 978-1-4020-3604-0

Published by Springer,
P.O. Box 17, 3300 AA Dordrecht, The Netherlands.

Sold and distributed in North, Central and South America
by Springer,
101 Philip Drive, Norwell MA 02061, U.S.A.

In all other countries, sold and distributed
by Springer,
P.O. Box 322, 3300 AH Dordrecht, The Netherlands.

Printed on acid-free paper

Printed in the Netherlands

To our mothers:
Renuka and Tajolmolouk

And our fathers:
Kula and Asghar

Table of Contents

Editorial

Multiphase flow in porous media is an extremely important process in a number of industrial and environmental applications, at various spatial and temporal scales. Thus, it is necessary to identify and understand multiphase flow and reactive transport processes at microscopic scale and to describe their manifestation at the macroscopic level (core or field scale). Current description of macroscopic multiphase flow behavior is based on an empirical extension of Darcy's law supplemented with capillary pressure-saturation-relative permeability relationships. However, these empirical models are not always sufficient to account fully for the physics of the flow, especially at scales larger than laboratory and in heterogeneous porous media. An improved description of physical processes and mathematical modeling of multiphase flow in porous media at various scales was the scope a workshop held at the Delft University of Technology, Delft, The Netherlands, 23–25 June, 2003. The workshop was sponsored by the European Science Foundation (ESF). This book contains a selection of papers presented at the workshop. They were all subject to a full peer-review process. A subset of these papers has been published in a special issue of the journal *Transport in Porous Media* (2005, Vol. 58, nos. 1–2).

The focus of this book is on the study of multiphase flow processes as they are manifested at various scales and on how the physical description at one scale can be used to obtain a physical description at a higher scale. Thus, some papers start at the pore scale and, mostly through pore-scale network modeling, obtain an average description of multiphase flow at the (laboratory or) core scale. It is found that, as a result of this upscaling, local-equilibrium processes may require a non-equilibrium description at higher scales. Some other papers start at the core scale where the medium is highly heterogeneous. Then, by means of upscaling techniques, an equivalent homogeneous description of the medium is obtained. A short description of the papers is given below.

Dahle, Celia, and Hassanizadeh present the simplest form of a pore-scale model, namely a bundle of tubes model. Despite their extremely simple nature, these models are able to mimic the major features of a porous medium. In fact, due to their simple construction, it is possible to reveal subscale mechanisms that are often obscured in more complex models. They use their model to demonstrate the pore-scale process that underlies dynamic capillary pressure effects.

Valvatne, Piri, Lopez and Blunt employ static pore-scale network models to obtain hydraulic properties relevant to single, two- and three-phase flow for a variety of rocks. The pore space is represented by a topologically disordered lattice of pores connected by throats that have angular cross sections. They consider single-phase flow of non-Newtonian as well as Newtonian fluids. They show that it is possible to use easily acquired data to estimate difficult-to-measure properties and to predict trends in data for different rock types or displacement sequences.

The choice of the geometry of the pore space in a pore-scale network model is very critical to the outcome of the model. In the paper by *Kainourgiakis, Kikkinides, Galani, Charlambopolous, and Stubos,* a procedure is developed for the reconstruction of the porous structure and the study of transport properties of the porous medium. The disordered structure of porous media, such as random sphere packing, Vycor glass, and North Sea chalk, is represented by three-dimensional binary images. Transport properties such as Kadusen diffusivity, molecular diffusivity, and permeability are determined through virtual (computational) experiments.

The pore-scale network model of Kainourgiakis *et al.* is employed by *Yiotis, Stubos, Boudouvis, Tsimpanogiannis, and Yortsos* to study drying processes in porous media. These include mass transfer by advection and diffusion in the gas phase, viscous flow in the liquid and gas phases, and capillary effects. Effects of films on the drying rates and phase distribution patterns are studied and it is shown that film flow is a major transport mechanism in the drying of porous materials.

Panfilov and Panfilova also start with a pore-scale description of two-phase flow, based on Washburn equation for flow in a tube. Subsequently, through a conceptual upscaling of the pore-scale equation, they develop a new continuum description of two-phase. In this formulation, in addition to the two fluid phases, a third continuum, representing the meniscus and called the M-continuum, is introduced. The properties of the M-continuum and its governing equations are obtained from the pore-scale description. The new model is analyzed for the case of one-dimensional flow. The remaining papers in this book regard upscaling from core scale and higher.

A procedure for upscaling dynamic two-phase flow in porous media is discussed by *Manthey, Hassanizadeh, and Helmig*. Starting with the Darcian description of two-phase flow in a (heterogeneous) porous medium, they perform fine-scale simulations and obtain macro-scale effective properties through averaging of numerical results. They focus on the study of an extended capillary pressure-saturation relationship that accounts for dynamic effects. They determine the value of the dynamic capillary pressure coefficient at various scales. They investigate the influence of averaging domain size, boundary conditions, and soil parameters on the dynamic coefficient.

The dynamic capillary pressure effect is also the focus of the paper by *Nieber, Dautov, Egorov, and Sheshukov*. They analyze a few alternative formulations of unsaturated flow that account for dynamic capillary pressure. Each of the alternative models is analyzed for flow characteristics under gravity-dominated conditions by using a traveling wave transformation for the model equations. It is shown that finger flow that has been observed during infiltration of water into a (partially) dry zone cannot be modeled by the classical Richard's equation. The introduction of dynamic effects, however, may result in unstable finger flow under certain conditions.

Nonequilibrium (dynamic) effects are also investigated in the paper by *Tavassoli, Zimmerman, and, Blunt*. They study counter-current imbibition, where the flow of a strongly wetting phase causes spontaneous flow of the nonwetting phase in the opposite direction. They employ an approximate analytical approach to derive an expression for a saturation profile for the case of non-negligible viscosity of the nonwetting phase. Their approach is particularly applicable to waterflooding of hydrocarbon reservoirs, or the displacement of NAPL by water.

In the paper by *Pickup, Stephen, Ma, Zhang and Clark*, a multistage upscaling approach is pursued. They recognize the fact that reservoirs are composed of a variety of rock types with heterogeneities at a number of distinct length scales. Thus, in order to upscale the effects of these heterogeneities, one may require a series of stages of upscaling, to go from small-scales (mm or cm) to field scale. They focus on the effects of steady-state upscaling for viscosity-dominated (water) flooding operations.

Gielen, Hassanizadeh, Leijnse, and Nordhaug present a dynamic pore-scale network model of two-phase flow, consisting of a three-dimensional network of tubes (pore throats) and spheres (pore bodies). The flow of two immiscible phases and displacement of fluid–fluid interface in the network is determined as a function of time using the Poiseuille flow equation. They employ their model to study dynamic effects in capillary pressure-saturation relationships and determine the value of the dynamic capillary pressure coefficient. As expected, they find a value that is one to two orders of magnitude larger than the value determined by Dahle *et al.* for a much simpler network model.

Eichel, Helmig, Neuweiler, and Cirpka present an upscaling method for two-phase in a heterogeneous porous medium. The approach is based on a percolation model and volume averaging method. Classical equations of two-phase flow are assumed to hold at the small (grid) scale. As a result of upscaling, the medium is replaced by an equivalent homogeneous porous medium. Effective properties are obtained through averaging results of fine-scale numerical simulations of the heterogeneous porous medium. They apply their upscaling technique to experimental data of a DNAPL infiltration experiment in a sand box with artificial sand lenses.

The editors wish to acknowledge an Exploratory Workshop Grant awarded by the European Science Foundation under its annual call for workshop funding in Engineering and Physical Sciences, which made it possible to organize the Workshop on Recent Advances in Multiphase Flow and Transport in Porous Media. We would like to express our sincere gratitude to colleagues who performed candid and valuable reviews of the original manuscripts. The publishing staffs of Springer are gratefully acknowledged for their enthusiasms and constant cooperation and help in bringing out this book.

The Editors

Dr. Diganta Bhusan Das, *Department of Engineering Science, The University of Oxford, Oxford OX1 3PJ, UK.*

Professor S.M. Hassanizadeh, *Department of Earth Sciences, Utrecht University, 3508 TA Utrecht, The Netherlands.*

Transp Porous Med (2005) 58:5–22
DOI 10.1007/s11242-004-5466-4

Bundle-of-Tubes Model for Calculating Dynamic Effects in the Capillary-Pressure-Saturation Relationship

HELGE K. DAHLE[1,*], MICHAEL A. CELIA[2] and
S. MAJID HASSANIZADEH[3]
[1] *Department of Mathematics, University of Bergen, Johns Brungst. 12, N-5008 Bergen, Norway*
[2] *Department of Civil and Environmental Engineering, Princeton University*
[3] *Department of Earth Sciences, Utrecht University*

(Received: 18 August 2003; in final form: 27 April 2004)

Abstract. Traditional two-phase flow models use an algebraic relationship between capillary pressure and saturation. This relationship is based on measurements made under static conditions. However, this static relationship is then used to model dynamic conditions, and evidence suggests that the assumption of equilibrium between capillary pressure and saturation may not be be justified. Extended capillary pressure–saturation relationships have been proposed that include an additional term accounting for dynamic effects. In the present work we study some of the underlying pore-scale physical mechanisms that give rise to this so-called dynamic effect. The study is carried out with the aid of a simple bundle-of-tubes model wherein the pore space of a porous medium is represented by a set of parallel tubes. We perform virtual two-phase flow experiments in which a wetting fluid is displaced by a non-wetting fluid. The dynamics of fluid–fluid interfaces are taken into account. From these experiments, we extract information about the overall system dynamics, and determine coefficients that are relevant to the dynamic capillary pressure description. We find dynamic coefficients in the range of $10^2 - 10^3\,\mathrm{kg\,m^{-1}\,s^{-1}}$, which is in the lower range of experimental observations. We then analyze certain behavior of the system in terms of dimensionless groups, and we observe scale dependency in the dynamic coefficient. Based on these results, we then speculate about possible scale effects and the significance of the dynamic term.

Key words: two-phase flow in porous media, dynamic capillary pressure, pore-scale network models, bundle-of-tubes, volume averaging

1. Introduction

Traditional equations that describe two-phase flow in porous media are based on conservation equations which are coupled to material-dependent

---

*Author for correspondence: e-mail: reshd@mi.uib.no

constitutive equations. One of the traditional constitutive equations is an algebraic relationship between capillary pressure, P_c (the difference between equilibrium phase pressures) and fluid phase saturation, S_α (the fraction of void space occupied by the fluid phase α). While this constitutive relationship is typically highly complex, including nonlinearity and hysteresis as well as residual phase saturations, it is nonetheless algebraic. The algebraic nature means that a change in one of the variables implies an instantaneous change in the other, such that the relationship between P_c and S is an equilibrium relationship. For an equilibrium relationship to be appropriate, the time scale of any dynamics associated with the processes that govern the relationship must be fast relative to the dynamics associated with other system processes. Time scales to reach equilibrium in laboratory experiments (Stephens, 1995) make this assumption questionable.

Recently, the relationship between P_c and S has been generalized, based on thermodynamic arguments by Gray and Hassanizadeh (see Hassanizadeh and Gray, 1990, 1993a, b; Gray and Hassanizadeh, 1991a, b). The extended relationship reads:

$$(p^{nw} - p^w) - P_c(S_w) = f\left(S_w, \frac{\partial S_w}{\partial t}\right), \tag{1}$$

where f denotes an unspecified function depending on saturation and its rate of change. Their contention is that this condition includes dynamic effects and is valid under unsteady state and nonequilibrium conditions. This kind of relationship has previously been considered by Stauffer (1978), and similar results occur in the classic book by Barenblatt et al. (1990), see also Silin and Patzek (2004). Dynamic effects may also occur as a consequence of upscaling of effective parameters in two-phase flow, see Bourgeat and Panfilov (1998). Recently, Hassanizadeh et al. (2002) analyzed experimental data sets from the literature and showed that dynamic effects are present in standard laboratory experiments to determine P_c as a function of S, although most laboratory experiments are designed to avoid dynamic effects by using small pressure increments. Hassanizadeh et al. (2002) and Dahle et al. (2002) also showed that this new relationship can easily be included in numerical simulations, and that effects on problems involving infiltrating fluid fronts could be significant, if the dynamic coefficient exhibits scale dependence.

In the present work, we consider some of the underlying physical mechanisms that give rise to this so-called dynamic effect. To do this, we analyze a simple bundle-of-tubes model that represents the pore space of a porous medium. This model is analogous to the recent model of Bartley and Ruth (1999, 2001), who used a bundle-of-tubes model to analyze dynamic effects in relative permeability, Bartley and Ruth (2001) also presented initial calculations on dynamic effects on the $P_c - S$ relationship.

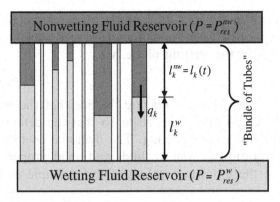

Figure 1. Bundle-of-tubes model.

In the model we present herein, we use a bundle-of-tubes model to analyze system behavior in the context of Figure 1. We perform virtual two-phase displacement experiments and mathematically track the dynamics of each fluid–fluid interface in two-fluid displacement experiments. From this we extract information about the overall system dynamics, and determine coefficients that are relevant to the dynamic description. We analyze certain behavior of the system in terms of dimensionless groups. Based on those results, we then speculate about possible scale effects and the significance of the dynamic term.

The paper is organized as follows. In the next section, we present background equations that are relevant to the derivations and calculations that follow. In the following section, we present the bundle-of-tubes model that is used to calculate system dynamics. We then describe the numerical experiments performed, and proceed to investigate certain scaling dependencies on the dynamic term. We end with a summary of the main findings and a discussion section.

2. Background Equations

The new relationship between P_c and S introduces a so-called dynamic capillary pressure, and hypothesizes that the rate of change of saturation is a function of the difference between the dynamic capillary pressure and the static, or equilibrium, capillary pressure. Assuming that a linear relationship holds, one will have, (Hassanizadeh and Gray, 1990):

$$-\tau \frac{\partial S_w}{\partial t} = P_c^{dyn} - P_c^{stat}(S_w).$$ (2)

In Equation (2), P_c^{stat} is the static or equilibrium capillary pressure, taken to be the capillary pressure that is traditionally measured in equilibrium pressure cell tests, see for example Stephens (1995); τ is a coefficient that we will call the 'dynamic coefficient'; and P_c^{dyn} is the dynamic capillary pressure, defined as the difference between the volume-averaged pressure in the nonwetting phase and that in the wetting phase, viz.

$$P_c^{dyn} = \langle p^{nw} \rangle - \langle p^w \rangle, \tag{3}$$

where the angular brackets imply volume averaging. Notice that the averaging procedure introduces a length (and time) scale, so that the definition of (3) will be linked to these scales of averaging. The dynamic coefficient may still be a function of saturation as well as fluids and solid properties. Stauffer (1978) has suggested the following scaling of the dynamic coefficient:

$$\tau = \frac{\phi \mu}{k} \frac{\alpha}{\lambda} \left(\frac{p^e}{\rho g} \right)^2, \tag{4}$$

where k is the intrinsic permeability, μ and ρ are the viscosity and density of the (wetting) fluid, g is the gravity constant, $\alpha = 0.1$ and λ, p^e are coefficients in the Brook–Corey formula.

Ideally, in order to investigate the validity of Equations (2) and (4), one should perform a large number of experiments, in which fluid pressures and saturation should be measured under a number of different conditions and for a variety of soil and fluid combinations. That, however, would be extremely costly and time consuming. At these early stages of research on dynamic capillary effects, it would be useful to carry out some theoretical work in order to gain insight into the various aspects of this phenomenon. Thus, in this paper, we try to gain insight into the underlying physics of Equation (2) and the effect of various soil and fluid properties on the value of τ. We carry out this work by studying fluid–fluid displacement at the pore scale within a simple pore-scale network model, composed of a bundle of capillary tubes. A schematic of the system is shown in Figure 1.

Consider a single capillary tube, with one end of the tube connected to a non-wetting-phase reservoir and the other end connected to a wetting-phase reservoir. The corresponding reservoir pressures are denoted by P_{res}^{nw} and P_{res}^w, respectively. Assume that both reservoir pressures may be controlled, and are set so that their difference is given by $\Delta P = P_{res}^{nw} - P_{res}^w$. If the tube has radius r, and is initially filled with wetting fluid, then non-wetting fluid will invade the tube if the pressure difference exceeds the displacement pressure given by the Young-Laplace criterion (Dullien, 1992) $\Delta P > 2\sigma^{wn} \cos\theta / r$, where σ^{wn} denotes interfacial tension between the wetting and non-wetting fluids, and θ is contact angle. Once this occurs, the fluid movement may be approximated by the Washburn equation (Washburn, 1921):

$$q = dl/dt = -\frac{r^2}{8\bar{\mu}(l)L}(-\Delta P + \bar{\rho}(l)Lg \sin \Phi + p_c(r)). \tag{5}$$

In Equation (5), $\bar{\mu}$ and $\bar{\rho}$ are length-averaged viscosity and density, respectively, of the fluids within the tube, $l = l(t)$ is the position of the interface in the tube of length L, Φ is the angle the tube makes with the vertical, and p_c is the local capillary pressure, taken to be equal to the displacement pressure,

$$p_c(r) = \frac{2\sigma^{wn}\cos\theta}{r}, \qquad (6)$$

To motivate the use of a bundle-of-tubes model, and to show the connection to the larger (continuum–porous-medium) scale, consider the following simple scaling argument. Assume Equations (5) and (6), applied to a large collection of pore tubes of different radii, govern the fluid flow through some portion of a porous medium. Then the analogies between the small-scale quantities in Equations (5) and (6), and those defined at the continuum-porous-medium scale, may be identified, under both static and dynamic conditions, as:

$$PS: \quad \frac{dl}{dt} = -\frac{r^2}{8\bar{\mu}(l)L} \qquad \underbrace{(-\Delta P + \bar{\rho}(l)Lg\sin\Phi}_{\downarrow} + \underbrace{p_c)}_{\downarrow}$$

$$CS: \quad -\frac{dS_w}{dt} = -\frac{1}{\tau} \qquad (\qquad -P_c^{dyn} \qquad + P_c^{stat})$$

Here PS denotes 'pore scale' and CS denotes 'continuum scale'. We see the direct correspondence between the dynamic displacement and the interface movement, and the associated upscaled versions of average phase pressure evolution and phase saturation changes. In particular, both $dl/dt = 0$ and $dS_w/dt = 0$ at equilibrium, although the units are different due to volume averaging. This provides motivation to use a bundle-of-tubes model to investigate more complex aspects of dynamic phase pressures, the associated dynamic capillary pressure, and its relationship to saturation dynamics. For more details on the use of these ideas in conjunction with pore-scale network models, we refer to Dahle and Celia (1999) and Hassanizadeh et al. (2002).

3. Bundle-of-Tubes Model

3.1. VOLUME AVERAGING

One of the main advantages of pore-scale network models is that variables that are difficult or impossible to measure physically can be computed directly from the network model. In the present case, we are interested in calculation of volume-averaged phase pressures, local and averaged capillary pressure, averaged phase saturations, and local interfacial velocities and associated changes in average phase saturations. To perform these calculations, we let V denote an averaging volume within the domain of the pore-scale network model, and introduce the indicator function γ defined by

$$\gamma_\alpha(\mathbf{x}, t) = \begin{cases} 1 \text{ if phase } \alpha \text{ at } (\mathbf{x}, t), \\ 0 \text{ otherwise.} \end{cases} \tag{7}$$

We then define

$$V_p = \iiint_V (\gamma_{nw} + \gamma_w) d\mathbf{x}, \quad V_w(t) = \iiint_V \gamma_w(\mathbf{x}, t) d\mathbf{x}, \tag{8}$$

and

$$V_{nw}(t) = V_p - V_w(t). \tag{9}$$

Here V_p is the total pore space of the averaging volume, $\phi = V_p/V$ is the porosity, and $V_\alpha(t)$ is the pore space occupied by phase α, with $\alpha = w$ for the wetting phase and $\alpha = nw$ for the non-wetting phase. Average state variables like saturation and phase pressures can now be defined as follows:

$$S_w(t) = \frac{V_w(t)}{V_p} = 1 - S_{nw}(t), \tag{10}$$

$$\langle p^\alpha \rangle = \frac{\iiint_V p_\alpha(\mathbf{x}, t)\gamma_\alpha(\mathbf{x}, t) d\mathbf{x}}{\iiint_V \gamma_\alpha(\mathbf{x}, t) d\mathbf{x}}, \quad \alpha = w, nw. \tag{11}$$

The bracket notation $\langle \rangle$ is used to denote average.

3.2. GEOMETRY OF THE BUNDLE-OF-TUBES MODEL

The bundle-of-tubes pore-scale model represents the pore space by a number, N, of non-intersecting capillary tubes. Each tube has length L, with one end of the tube connected to a reservoir of nonwetting fluid and the other end connected to a reservoir of wetting fluid (see Figure 1). Each tube is assigned a different radius r, with the radii drawn from a cut-off log-normal distribution

$$f(r; \sigma_{nd}) = \frac{\sqrt{2}\exp\left[-\frac{1}{2}\left(\frac{\ln\frac{r}{r_{ch}}}{\sigma_{nd}}\right)^2\right]}{\sqrt{\pi\sigma_{nd}^2}\, r\left[\text{erf}\left(\frac{\ln\frac{r_{max}}{r_{ch}}}{\sqrt{2\sigma_{nd}^2}}\right) - \text{erf}\left(\frac{\ln\frac{r_{min}}{r_{ch}}}{\sqrt{2\sigma_{nd}^2}}\right)\right]}. \tag{12}$$

Here r_{ch} and σ_{nd} are the mean and variance of the parent distribution. We have conveniently fixed the maximum and minimum radius to be

$r_{max} = 10^2 r_{ch}$ and $r_{min} = 10^{-3} r_{ch}$. Following Dullien (1992), let $V = L^3$ be the averaging volume of the bundle, and define the average of the pth power of r_k by:

$$\langle r^p \rangle = \sum_{k=1}^{N} r_k^p / N. \tag{13}$$

Then the porosity is given by

$$\phi = \frac{V_p}{L^3} = \frac{\pi N \langle r^2 \rangle}{L^2}, \text{ or } L = \left(\frac{\pi N \langle r^2 \rangle}{\phi} \right)^{1/2}, \tag{14}$$

In our computations we will specify the porosity ϕ and calculate the length of the tubes L from this formula. From the parallel tubes model, we may calculate an intrinsic permeability, k, for the bundle as

$$\left. \begin{aligned} Q &= \sum_k \frac{\pi r_k^4}{8\mu} \frac{\Delta P}{L} = \frac{\pi N \langle r^4 \rangle}{8\mu} \frac{\Delta P}{L} \\ Q &= \frac{kL^2}{\mu} \frac{\Delta P}{L} \end{aligned} \right\} \Rightarrow k = \frac{\phi \langle r^4 \rangle}{8 \langle r^2 \rangle}, \tag{15}$$

where we have used Equation (14).

3.3. COMPUTATIONAL ALGORITHM

Assume that the tubes are ordered by decreasing radius such that $r_k \geqslant r_{k+1}$, $k = 1, 2, \ldots, N-1$, and that they are initially filled by wetting fluid. The bundle is then drained by gradually increasing the non-wetting reservoir pressure P_{res}^{nw}, while the wetting reservoir pressure, P_{res}^{w}, is kept fixed, say equal to zero. The dynamics of each interface is assumed to be governed by Equation (5). However, in order to save on algebra, the gravity will be neglected in the following analysis and the two fluids are assumed to have the same viscosity μ, leading to a pressure distribution within the tube as shown in Figure 2. Thus, once the non-wetting reservoir pressure exceeds the displacement pressure of tube k, the location of that interface at any time t, $l = l_k(t)$, is given by,

$$l_k(t) = q_k \cdot (t - t_0) + l_k^0, \tag{16}$$

where

$$q_k = -\frac{r_k^2}{8\mu L} (-\Delta P + p_c(r_k)), \tag{17}$$

and l_k^0 is the position of the interface at time t_0. When the interface reaches the wetting reservoir, $l_k = L$, that interface will be considered to be trapped, with $q_k = 0$, and the pressure in the corresponding drained tubes is kept

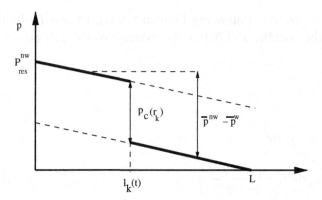

Figure 2. Pressure distribution in a single tube containing two fluids of equal viscosity separated by an interface located at $l = l_k(t)$.

constant at P_{res}^{nw}. By averaging we obtain the following expression for the saturation of the wetting phase at any given time t:

$$S_w(t) = \frac{V_w(t)}{V_p} = 1 - \frac{\sum_k \pi r_k^2 l_k}{V_p}.$$ (18)

The time derivative of this saturation is

$$\frac{dS_w}{dt} = -\frac{1}{V_p} \sum_k \pi r_k^2 \frac{dl_k}{dt} = -\frac{1}{V_p} \sum_k \pi r_k^2 q_k.$$ (19)

By using Equation (11) we obtain the average phase pressures ($\alpha = w, nw$):

$$<p^\alpha> = \frac{1}{V_\alpha(t)} \sum_k \pi r_k^2 l_k^\alpha (\pm \frac{1}{2} \Delta p_k^\alpha + P_{res}^\alpha),$$ (20)

where

$$\Delta p_k^\alpha = \begin{cases} \frac{l_k^\alpha}{L}(-\Delta P + p_c(r_k)), & 0 < l_k^\alpha < L; \\ 0 & l_k^\alpha = L. \end{cases}$$ (21)

Here $l_k^{nw} = l_k(t)$, $l_k^w = L - l_k(t)$ and the plus sign is chosen if $\alpha = nw$. These phase pressures are then used in Equation (3) to define the dynamic capillary pressure. At equilibrium the capillary pressure over an interface has to exactly balance the boundary pressures. This leads to the following definition of a static capillary pressure:

$$P_c^{stat}(S_w) = p_c(r_k), \quad S_w^{k-1} \leq S_w \leq S_w^k \text{ with } S_w^k = 1 - \frac{\sum_i^{k-1} \pi r_i^2 L}{V_p}.$$ (22)

Note that P_c^{stat} is defined stepwise as the displacement pressure of successive tubes. In Figure 3 dynamic and static capillary-pressure–saturation

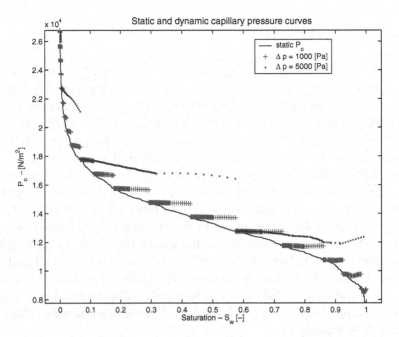

Figure 3. Dynamic and static capillary-pressures–saturation curves.

relationships are compared for two different drainage experiments. The only difference between these experiments is that different pressure increments, Δp_{step}, are used to update the nonwetting reservoir pressure P_{res}^{nw}.

Observe that the dynamic capillary-pressure curves in Figure 3 are always above the static curve, which is consistent with the theory leading to Equation (2). Another interesting feature of this Figure is the non monotonicity of the dynamic P_c-curve for large saturation. Similar behavior has also been observed in dynamic network simulations, e.g. Hassanizadeh *et al.* (2002). To explain the behavior in Figure 3, consider a single tube, k, with a moving interface at $l = l_k(t)$. Since the viscosities of the fluids are equal, the pressure gradient has to be equal within each fluid phase of the tube, see Figure 2, and the average phase pressures in that tube are given by:

$$\bar{p}_k^{nw} = P_{res}^{nw} - \frac{l_k}{2L}(P_{res}^{nw} - p_c(r_k)), \quad \bar{p}_k^{w} = \frac{L - l_k}{2L}(P_{res}^{nw} - p_c(r_k)). \tag{23}$$

Thus, the average phase pressures in a single tube will decrease at the same rate, whereas the difference is constant in time:

$$\bar{p}_k^{nw} - \bar{p}_k^{w} = \frac{1}{2}(P_{res}^{nw} + p_c(r_k)). \tag{24}$$

If we consider the ensemble of tubes, the average phase pressures, Equation (20), may alternatively be written:

$$\langle p^\alpha \rangle = \frac{1}{V_\alpha(t)} \sum_k \bar{p}_k^\alpha V_\alpha^k, \tag{25}$$

where V_α^k is the volume occupied by phase α in tube k, $\bar{p}_k^{nw} = P_{res}^{nw}$ if the interface in tube k is trapped at $l_k = L$, and $\bar{p}_k^w = 0$ if the interface is trapped at $l_k = 0$. At high saturations we may assume that all the non-wetting fluid is associated with moving interfaces. Since the flow rate in each tube is constant, all the volumes associated with the non-wetting fluids are then changing proportional to time t. It follows that $\langle p^{nw} \rangle$ has to be decreasing function of time (i.e. decreasing saturation), since all the weights, $\bar{p}_k^{nw}$, are decreasing. At the point in time when interfaces starts to get trapped at the outflow boundary, the associated weights will increase, and $\langle p^{nw} \rangle$ may start to increase in time. On the other hand, for high saturations, the volumes occupied by the wetting fluid is mainly associated with interfaces that are immobile at the inflow boundary giving weights $\bar{p}_k^w = 0$, so that $\langle p^w \rangle \approx 0$. Thus, $P_c^{dyn} \approx \langle p^{nw} \rangle$ has to be a decreasing function with time in this case. By looking at Figure 3, this behavior is clearly apparent for $0.9 < S_w < 1$ and $\Delta p_{step} = 5000 Pa$. For $S_w \approx 0.9$ a sufficient number of interfaces become trapped at the outflow boundary, leading to a change of slope in the dynamic $P_c - S$ curve.

4. Numerical Experiments

In the numerical tests reported herein, a set of radii are generated based on the log-normal distribution, and these radii define one realization of the pore-scale geometry. For a given realization, the tubes are drained by imposition of step-wise changes in pressure in the nonwetting reservoir. Initially we choose $P_{res}^{nw} = p_c(r_1) + \Delta p_{step}$ and then increase P_{res}^{nw} subsequently by Δp_{step} each time an equilibrium is reached (meaning that no further interfaces will move). In this way the entire bundle is drained, and we can compute $P_c^{stat} - P_c^{dyn}$ and dS_w/dt at a given set of target saturations $S_{target} \in \{0.1, 0.2, \ldots, 0.9\}$. To obtain a sufficiently large number of data points at each target saturation we vary the pressure step according to

$$\Delta p_{step} = n \cdot \delta p, \quad n = 1, 2, \ldots, N_{step}, \quad \text{with } \delta p = (1.1 p_c(r_N) - p_c(r_1))/N_{step}.$$

Observe that the largest pressure increment is chosen such that the bundle will drain in a single step. We have chosen $N_{step} = 50$, and if nothing else is specified other parameters for the bundle are chosen as listed in Table I.

In Figure 4, $P_c^{stat} - P_c^{dyn}$ is plotted against dS_w/dt at target saturations 0.2, 0.5 and 0.8. Observe that the data points appear to behave linearly somewhat away from the origin, while close to the origin we have that $P_c^{dyn} \to P_c^{stat}$ as $dS_w/dt \to 0$ in a nonlinear fashion. We may fit a straight line through the linear portion of the curve, with parameters τ and β defined as slope and intercept,

Table I. Parameters for bundle of tube model. Length L of tubes and intrinsic permeability k are calculated from one realization of the bundle using Equations (14) and (15)

Parameter	Description	Value
N	Number of tubes	1000
N_{step}	Number of pressure increments	50
r_{ch}	Mean value pore-size distribution	10^{-5} [m]
r_{min}	Lower cut-off radius	$10^{-3}r_{ch}$
r_{max}	Upper cut-off radius	$10^{2}r_{ch}$
σ_{nd}	Variance of pore-size distribution	0.2
μ	Viscosity	0.5×10^{-2} [kg m^{-1} s^{-1}]
σ^{wn}	Surface tension	7.2×10^{-2} [kg s^{-2}]
θ	Contact angle	0 (radians)
ϕ	Porosity	0.3
L	Length	$\sim 10^{-3}$ [m]
k	Permeability	$\sim 4.8 \times 10^{-12}$ [m^2]
τ	Dynamic coefficient	~ 274 [kg m^{-1} s^{-1}]
β	Intercept	$\sim 1.5 \times 10^{3}$ [kg m^{-1} s^{-2}]

Similarly, the dynamic coefficient τ and the intercept β is calculated from the same realization at saturation $S_w = 0.5$

$$-\tau \partial S_w/\partial t + \beta = P_c^{\text{dyn}} - P_c^{\text{stat}}, \qquad (26)$$

where $\tau > 0$, $\beta > 0$ may be functions of S_w and other parameters. Based on Stauffer's formula (4) we may conjecture that

$$\tau k/\phi \mu L^2 = \Pi_\tau(S_w, \sigma_{nd}). \qquad (27)$$

Here L should be interpreted as a characteristic length scale associated with the averaging volume. We also conjecture that

$$\beta/\sigma_{nd} P_c^{ch} = \Pi_\beta(S_w), \qquad (28)$$

where $P_c^{ch} = 2\sigma^{wn} \cos\theta/r_{ch}$ and r_{ch} is the mean of the pore size distribution.

To determine values of the parameters τ and β, and to test the conjectures put forth in Equations (27) and (28), we run a series of numerical experiments and analyze the results. As part of this analysis, we determine a regression line through the linear part of the plots (see for example Figure 4). To compute the regression line in a systematic manner, the data points are first normalized to fall within the interval $[-1, 0]$. A regression line is then calculated for all data points associated with $dS_w/dt < -0.3$ on the normalized plot. The regression line is then transformed back to the original coordinate system. The slope of the line gives the estimate for τ while the intercept gives β. Note that $\beta \neq 0$ corresponds to existence of a

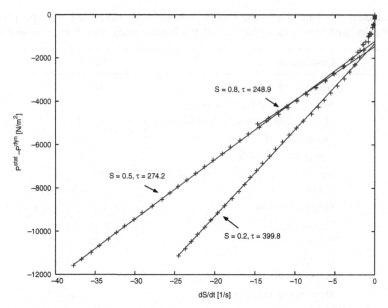

Figure 4. $P_c^{stat} - P_c^{dyn}$ versus dS_w/dt at saturations $S_w = 0.2, 0.5, 0.8$.

nonlinear region near the origin. The magnitude of β reflects the degree of this nonlinearity. In all our simulations, the slope of the regression line has been positive and the *curvature* of the data points have been such that the vertical axis intersection has been below the origin.

The proposed conjectures can now be tested by systematically varying the parameters associated with our bundle-of-tubes model. For each new value of a specified parameter, a new realization of the bundle is generated and this bundle is then drained using the N_{step} different pressure steps to obtain regression lines as in Figure 4. The parameters that are varied are N (number of tubes), ϕ (porosity), μ (viscosity), r_{ch} (mean pore-size distribution), σ_{nd} (variance of pore-size distribution), and θ (contact angle). Note that varying θ is equivalent to varying the surface tension σ^{wn}. It is also

Table II. Results from varying different parameters, keeping the others fixed as in Table I

Parameter	Range	k	L	τ	β
N	200–10,000	4.77×10^{-12}	$N^{1/2}$	N	*indep.*
ϕ	0.05–0.45	ϕ	$\phi^{-1/2}$	$\phi^{-1/2}$	*indep.*
r_{ch}	$10^{-6} - 10^{-4}$	r_{ch}^2	r_{ch}	*indep.*	r_{ch}^{-1}
μ	$10^{-4} - 10^{-1}$	4.79×10^{-12}	0.11×10^{-2}	μ	*indep.*
θ	$0 - 1.5608$	4.79×10^{-12}	0.11×10^{-2}	*indep.*	$\cos\theta$
σ_{nd}	$0.1 - 0.6$	*	*	*	σ_{nd}

The symbol * means that no obvious power law was found.

possible to vary the lower- and upper-cut-off radius r_{min} and r_{max} independently. However, for this study they are kept constant with values as given in Table I. The findings of our numerical simulations are summarized in Table II. For example, the number of tubes is increased from $N = 200$ to $N = 10000$ with step size 200 tubes. As expected, we observe that the permeability $k \sim 4.77 \times 10^{-12}[m^2]$ is essentially constant, i.e: k varies randomly around a mean value of $4.77 \times 10^{-12}[m^2]$ for various realizations of the bundle. Furthermore, $L \sim N^{1/2}$, and $\tau \sim N$, whereas β is essentially independent of N as $N = 200, 400, \ldots, 10,000$. Similar results are tabulated when varying the other parameters, see Table II. However, it turns out that the variance of the pore-size distribution σ_{nd}, is a special parameter. We let σ_{nd} vary linearly between $\sigma_{nd} = 0.1$ and $\sigma_{nd} = 0.6$ using 50 steps. Both k and L increase with σ_{nd} but no obvious power law dependency is found. Similarly, we find no obvious dependency with respect to τ and σ_{nd}. In fact, τ-values for smaller saturations increase with respect to σ_{nd} whereas they decrease at the larger saturation values. On the other hand it appears that $\beta \sim \sigma_{nd}$, although the fluctuations in the data points are fairly large for the larger values of σ_{nd}.

For each parameter that is varied, we have plotted the mean value for the dimensional groupings Π_τ and Π_β at the specified target saturations, see Figures 5–7. The error bars in these plots give the variance of the fluctuations around the mean value, due to different realizations of the

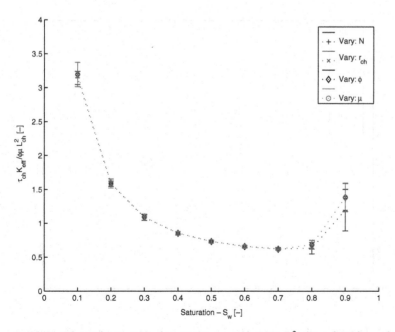

Figure 5. Dimensional grouping $\Pi_\tau(S_w, \sigma_{nd}) = \tau k/\phi \mu L^2$ as a function of saturation is fixed at $\sigma_{nd} = 0.2$. Variance of the pore-size distribution is fixed at $\sigma_{nd} = 0.2$.

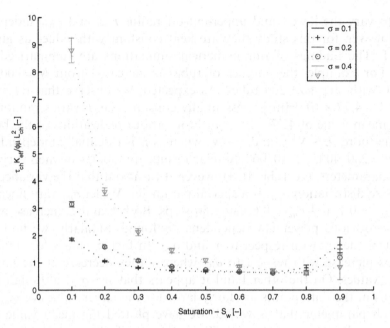

Figure 6. Dimensional grouping $\Pi_\tau(S_w, \sigma_{nd}) = \tau k/\phi \mu L^2$ as a function of saturation. Each curve represents a different variance of the pore-size distribution: $\sigma_{nd} =$
0.1, 0.2, 0.4.

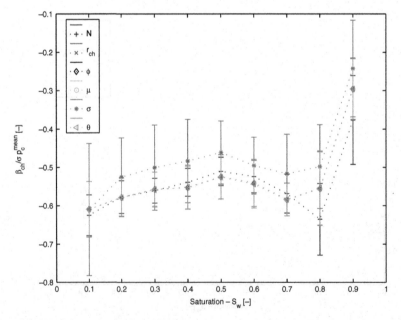

Figure 7. Dimensional grouping $\Pi_\beta(S_w) = \beta/\sigma_{nd} P_c^{ch}$ as a function of saturation.

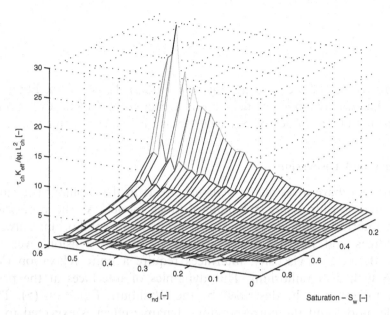

Figure 8. Dimensional grouping $\Pi_\tau(S_w, \sigma_{nd}) = \tau k/\phi\mu L^2$ as a function of saturation and variance of the pore-size distribution.

bundle for each update of the a specific parameter. We observe that Figures 5 and 7 reconcile the parameter dependencies of Π_τ and Π_β fairly well. In Figure 8, we do not include the data related to varying the variance of the pore-size distribution, simply because we are not able to make this parameter fit into the dimensional grouping of Π_τ. In Figure 6, we have plotted the dimensional grouping Π_τ when the number of tubes N is varied from $N = 200$ to $N = 10,000$, and for three different values of σ_{nd}. This Figure illustrates the difficulty associated with the parameter σ_{nd}. We are simply not able to include σ_{nd} into the dimensional grouping Π_τ to make this independent of σ_{nd}, because the dependency of this parameter is coupled to the saturations. We therefore suggest that $\Pi_\tau = \Pi_\tau(S_w, \sigma_{nd})$. This surface is plotted in Figure 8. A possible explanation for the more complicated dependency on σ_{nd} is related to the observation that $\tau \sim k^{-1}$. When σ_{nd} is increased we get more tubes with smaller and larger radius. This means that when we estimate τ for larger saturations the 'local' permeability over that section of the bundle must increase with σ_{nd}. Since τ is inversely proportional to permeability we should therefor expect τ to decrease for larger saturations when σ_{nd} is increased. On the other hand for smaller saturations the 'local' permeability should decrease with σ_{nd} resulting in an increase in τ.

Finally, by Equation (27), we have that

$$\tau(S_w) = \frac{\phi\mu L^2}{k} = \Pi_\tau(S_w),\tag{29}$$

for a fixed variance of the pore-size distribution σ_{nd}. Hence, from Figure 5, it follows that the dynamic coefficient τ is a decreasing function of saturation except for larger values of S_w, where τ is an increasing function.

5. Summary and Discussion

In this paper we have investigated *dynamic* effects in the capillary pressure–saturation relationship using a bundle-of-tubes model. At the pore-scale, fluid–fluid interfaces will always move to produce an equilibrium between external forces and internal forces created by surface tension over the interfaces. Because of viscosity, interfaces require a finite relaxation time to achieve such an equilibrium. This dynamics of interfaces at the pore-scale may for example be described by the Washburn Equation (5). This is a simple model and the corresponding dynamic effect is expected to be small. The calculated value of τ ($\sim 274\,\mathrm{kg\,m^{-1}\,s^{-1}}$) is indeed very small. For a more complicated pore-scale network model, larger values for τ are obtained. For example, for a three-dimensional pore-scale network model Gielen *et al.* (2004) obtained values of order $10^4 - 10^5\,\mathrm{kg\,m^{-1}\,s^{-1}}$. When micro-scale soil heterogeneities are taken into account, even larger values for τ are found. For example, experimental results reported by Manthey *et al.* (2004) on a 6-cm long homogeneous soil sample yield a τ-value of about $10^5\,\mathrm{kg\,m^{-1}\,s^{-1}}$. At even larger scales, dynamics of interfaces must be associated with the time scale of changes in phase saturations.

Our analysis of the bundle-of-tubes model leads to the relationship (26) involving a dynamic coefficient τ and an intercept of the vertical axis β. We have investigated dimensionless groupings (27) and (28) containing τ and β, respectively. The dimensionless grouping involving τ shows a clear dependency on saturation, in particular for larger values of the variance of the pore-size distribution. It also shows that the dynamic coefficient τ increases as the square of the length scale L associated with the averaging volume. This suggest that the dynamic coefficient may become arbitrarily large as the averaging volume increases in size. However, we suspect that the length scale has to be tied to typical length scales associated with the problem under consideration, e.g. length scales associated with moving fronts, and not necessarily the length scale of the averaging volumes. We will investigate the dependency of τ with respect to typical length scales in future work. The dynamic effect observed in our bundle-of-tubes model is only due to the motion of single interfaces. The effect would have been larger if effects such as hysteresis in contact angle would have been included; e.g. a smaller contact angle during drainage compared to when the interface is at rest.

The relationship (26) may not be valid for small temporal changes in saturation due to the nonlinearity introduced by local capillary pressure. The magnitude of this nonlinearity is reflected in the size of the vertical axis intercept β. In fact, the dimensionless grouping involving β shows that this intercept is proportional to surface tension and contact angle of the fluid–fluid interface. On the other hand, the dimensionless grouping that contains β does not show any clear dependency on the saturation. If this turns out to be the case, the β-term may have no importance with respect to continuum scale models.

Acknowledgements

Partial support for this work was provided to H.K. Dahle by the Norwegian Research Council and Norsk Hydro under Grants 151400/210 and 450196, to M.A. Celia by the National Science Foundation under Grant EAR-0309607, and the research by S.M. Hassanizadeh has been carried out in the framework of project no. NOW/ALW 809.62.012 financed by the Dutch Organization for Scientific Research.

References

Barenblatt, G. I., Entov, V. M. and Ryzhik, V. M.: 1990, *Theory of Fluids Through Natural Rocks*, Kluwer Academic Publishing: Dordrecht.

Bartley, J. T. and Ruth, D. W.: 1999, Relative permeability analysis of tube bundle models, *Transport Porous Media* **36**, 161–187.

Bartley, J. T. and Ruth, D. W.: 2001, Relative permeability analysis of tube bundle models, including capillary pressure, *Transport Porous Media* **45**, 447–480.

Bourgeat, A. and Panfilov, M.: 1998, Effective two-phase flow through highly heterogeneous porous media: capillary nonequilibrium effects, *Comput. Geosci.* **2**, 191–215.

Dahle, H. K. and Celia, M.: 1999, A dynamic network model for two-phase immiscible flow, *Comput. Geosci.* **3**, 1–22.

Dahle, H. K., Celia, M. A., Hassanizadeh, S. M. and Karlsen, K. H.: 2002, A total pressure–saturation formulation of two-phase flow incorporating dynamic effects in the capillary-pressure–saturation relationship. In: *Proc. 14th Int. Conf. on Comp. Meth. in Water Resources, Delft, The Netherlands, June 2002*.

Dullien, A.: 1992, *Porous Media: Fluid Transport and Pore Structure*, Academic Press, 2nd edn. New York.

Gielen, T., Hassanizadeh, S. M., Leijnse, A., and Nordhaug, H. F.: 2004, *Dynamic Effects in Multiphase Flow: A Pore-Scale Network Approach*. Kluwer Academic Publisher.

Gray, W. and Hassanizadeh, S.: 1991a, Paradoxes and Realities in Unsaturated Flow Theory, *Water Resour. Res.* **27**(8), 1847–1854.

Gray, W. and Hassanizadeh, S.: 1991b, Unsaturated flow theory including interfacial phenomena. *Water Resour. Res.* **27**(8), 1855–1863.

Hassanizadeh, S., Celia, M. and Dahle, H.: 2002, Dynamic effects in the capillary pressure–saturation relationship and its impacts on unsaturated flow, *Vadose Zone J.* **1**, 38–57.

Hassanizadeh, S. and Gray, W.: 1990, Mechanics and thermodynamics of multiphase flow in porous media including interface boundaries, *Adv. Water Res.* **13**(4), 169–186.

Hassanizadeh, S. and Gray, W.: 1993a, Thermodynamic basis of capillary pressure in porous media, *Water Resour. Res.* **29**(10), 3389–3405.

Hassanizadeh, S. and Gray, W.: 1993b, Toward an improved description of the physics of two-phase flow, *Adv. Water Res.* **16**, 53–67.

Manthey, S., Hassanizadeh, S. M., Oung, O. and Helmig, R.: 2004, Dynamic capillary pressure effects in two-phase flow through heterogeneous porous media, In: *Proc. 15th Int. Conf. on Comp. Meth. in Water.*

Silin, D. and Patzek, T.: 2004, On Barenblatt's model of spontaneous countercurrent imbibition, *Transport Porous Media* **54**, 297–322.

Stauffer, F.: 1978, Time dependence of the relationships between capillary pressure, water content and conductivity during drainage of porous media. In: *IAHR Symp. On Scale Effects in Porous Media*, IAHR, (Madrid, Spain).

Stephens, D.: 1995, *Vadose Zone Hydrology*. Lewis Publ., Boca Raton, Florida.

Washburn, E.: 1921, The dynamics of capillary flow, *Phys. Rev.* **17**, 273–283.

Transp Porous Med (2005) 58:23–41
DOI 10.1007/s11242-004-5468-2

Predictive Pore-Scale Modeling of Single and Multiphase Flow

PER H. VALVATNE, MOHAMMAD PIRI, XAVIER LOPEZ and
MARTIN J. BLUNT*
Department of Earth Science and Engineering, Imperial College London, SW7 2AZ, U.K.

(Received: 15 August 2003; in final form: 25 February 2004)

Abstract. We show how to predict flow properties for a variety of rocks using pore-scale modeling with geologically realistic networks. The pore space is represented by a topologically disordered lattice of pores connected by throats that have angular cross-sections. We successfully predict single-phase non-Newtonian rheology, and two and three-phase relative permeability for water-wet media. The pore size distribution of the network can be tuned to match capillary pressure data when a network representation of the system of interest is unavailable. The aim of this work is not simply to match experiments, but to use easily acquired data to estimate difficult to measure properties and to predict trends in data for different rock types or displacement sequences.

Key words: multiphase flow, pore-scale modeling, relative permeability

1. Introduction

In network modeling the void space of a rock is represented at the microscopic scale by a lattice of pores connected by throats. By applying rules that govern the transport and arrangement of fluids in pores and throats, macroscopic properties, for instance capillary pressure or relative permeability, can then be estimated across the network, which typically consists of several thousand pores and throats representing a rock sample of a few millimeters cubed.

Until recently most networks were based on a regular lattice. The coordination number can vary depending on the chosen lattice (e.g. 5 for a honeycombed lattice or 6 for a regular cubic lattice). As has been noted by many authors (Chatzis and Dullien, 1997; Wilkinson and Willemsen, 1983) the coordination number will influence the flow behavior significantly. In order to match the coordination number of a given rock sample, which typically is between 3 and 8 (Jerauld and Salter, 1990), it is possible to remove throats at random from a regular lattice (Dixit *et al.*, 1997,

---

*Author for correspondence: e-mail: m.blunt@imperial.ac.uk

1999), hence reducing the connectivity. By adjusting the size distributions to match capillary pressure data, good predictions of absolute and relative permeabilities have been reported for unsaturated soils (Fischer and Celia, 1999; Vogel, 2000).

All these models are, however, still based on a regular topology that does not reflect the random nature of real porous rock. The use of networks derived from a real porous medium was pioneered by Bryant *et al.* They extracted their networks from a random close packing of equally-sized spheres where all sphere coordinates had been measured (Bryant and Blunt, 1992; Bryant *et al.*, 1993a, b). Predictions of relative permeability, electrical conductivity and capillary pressure were compared successfully with experimental results from sand packs, bead packs and a simple sandstone. Øren and coworkers at Statoil have extended this approach to a wider range of sedimentary rocks (Bakke and Øren, 1997; Øren *et al.*, 1998). It is usually necessary to create first a three-dimensional voxel based representation of the pore space that should capture the statistics of the real rock. This can be generated directly using X-ray microtomography (Dunsmuir *et al.*, 1991; Spanne *et al.*, 1994), where the rock is imaged at resolutions of around a few microns, or by using a numerical reconstruction technique (Adler and Thovert, 1998; Øren and Bakke, 2002). From this voxel representation an equivalent network (in terms of volume, throat radii, clay content etc) can then be extracted (Delerue and Perrier, 2002; Øren and Bakke, 2002). Using these realistic networks experimental data have been successfully predicted for Bentheimer (Øren *et al.*, 1998) and Berea sandstones (Blunt *et al.*, 2002).

2. Network Model

We use a capillary dominated network model that broadly follows the work of Øren, Patzek and coworkers (Øren *et al.*, 1998; Patzek, 2001). The extensions to three-phase flow are described by Piri and Blunt (2002). Incorporation of non-Newtonian flow is discussed in Lopez *et al.* (2003). Further details, including relevant equations, can be found in Blunt (1998), Øren *et al.* (1998) and Patzek (2001). The model simulates primary drainage, wettability alteration and any subsequent cycle of water flooding, secondary drainage and gas injection.

2.1. DESCRIPTION OF THE PORE SPACE

A three-dimensional voxel representation of either Berea sandstone or a sand pack (Table I) is the basis for the networks used in this paper. The pore space image is generated by simulating the random close packing of spheres of different size followed (in the case of Berea) by compaction,

Table I. Properties of the two networks used in this paper

Network	ϕ	K(D)	Pore radius range $(10^{-6}$ m)	Throat radius range $(10^{-6}$ m)	Average coordination number
Sand pack	0.34	101.8	3.2–105.9	0.5–86.6	5.46
Berea	0.18	3.148	3.6–73.5	0.9–56.9	4.19

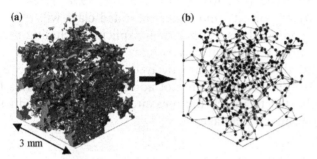

Figure 1. (a) A three-dimensional image of a sandstone with (b) a topologically equivalent network representation (Bakke and Øren, 1997; Øren *et al.*, 1998)

diagenesis and clay deposition. A topologically equivalent network of pores and throats is then generated with properties (radius, volume etc.) extracted from the original voxel representation, shown schematically in Figure 1. The networks were provided by other authors (Bakke and Øren, 1997; Øren *et al.*, 1998) – in this work we simply used them as input to our modeling studies. The Berea network represented a sample 3 mm cubed with 12,000 pores and 26,000 throats while the sand pack network contained 3,500 pores and 10,000 throats. With this relatively small number of elements, a displacement sequence can be run using standard computing resources in under a minute.

The cross-sectional shape of the network elements (pores or throats) is a circle, square or triangle with the same shape factor, $\Gamma = A/L^2$, as the voxel representation, where A is the cross sectional area and L the perimeter length. As the pore space becomes more irregular the shape factor decreases. Compared to the voxel image, the network elements are obviously only idealized representations. However, by maintaining the measured shape factor a quantitative measure of the irregular pore space is maintained. Fairly smooth pores with high shape factors will be represented by network elements with circular cross-sections, whereas more irregular pore shapes will be represented by triangular cross-sections, possibly with very sharp corners.

Using square or triangular shaped network elements allows for the explicit modeling of wetting layers where non-wetting phase occupies the center of the element and wetting phase remains in the corners. The pore space in real rock is highly irregular with wetting fluid remaining in grooves and crevices after drainage due to capillary forces. The wetting layers may only be a few microns in thickness, with little effect on the overall saturation or flow, but their contribution to wetting phase connectivity is of vital importance, ensuring low residual wetting phase saturation by preventing trapping (see, for instance, Blunt, 1998; Øren et al., 1998; Patzek, 2001). Micro-porosity and water saturated clays will typically not be drained during core analysis. Rather than explicitly including this in the network representation, a constant clay volume is associated with each element. The pore and throat shapes are derived directly from the pore space representation. In this work they are not adjusted to match data. The clay volume can be adjusted to match the measured connate or irreducible water saturation after primary drainage.

3. Single-Phase Non-Newtonian Flow

There are many circumstances where non-Newtonian fluids, particularly polymers, are injected into porous media, such as for water control in oil wells or to enhance oil recovery. In this section we will predict the single-phase properties of shear-thinning fluids in a porous medium from the bulk rheology. Several authors (see, for instance, Sorbie, 1991) have derived expressions to define an apparent shear rate experienced by the fluid in the porous medium from the Darcy velocity. In practice, apparent viscosity (μ_{app}) and Darcy velocity (q) are often the measured quantities. Experimental results suggest that the overall shape of the $\mu_{app}(q)$ curve is similar to that in the bulk $\mu(\gamma)$, where γ is the shear rate. Using dimensional analysis there is a length that relates velocity to shear rate. Physically this length is related to the pore size. One estimate of this length is the square root of the absolute permeability times the porosity, $K\phi$ (Sorbie, 1991). This allows the determination of in situ rheograms from the bulk measured $\mu(\gamma) : \mu_{app}(q) = \mu(\gamma = q/\sqrt{K\phi})$. Many authors have remarked that this method leads to in situ rheograms that are shifted from the bulk curve by a constant factor, α (Sorbie, 1991; Pearson and Tardy, 2002):

$$\mu_{app}(q) = \mu\left(\gamma = \alpha q/\sqrt{K\phi}\right) \tag{1}$$

Reported values for α vary depending on the approach chosen, but experimental results suggest it generally lies in the range 1 to 15. Pearson and Tardy (2002) reviewed the different mathematical approaches used to describe non-Newtonian flow in porous media. They concluded that none of the present continuum models give accurate estimates of bulk rheology and the pore structure and currently there is no theory that can predict its value reliably.

We will consider polymer solutions – representing Xanthan – whose bulk rheology is well-described using a truncated power-law:

$$\mu_{eff} = Max\left[\mu_{\infty}, Min\left(C\gamma^{n-1}, \mu_0\right)\right] \qquad (2)$$

where C is a constant and n is a power-law exponent. We can solve analytically for the relationship between pressure drop and flow rate for a truncated power-law fluid in a circular cylinder (Lopez et al., 2003). Our network models are, however, mainly composed of irregular triangular-shaped pores and throats. To account for non-circular pore shapes we replace the inscribed radius of the cylinder R in the relationship between flow rate and pressure drop with an appropriately defined equivalent radius, R_{equ}. We use an empirical approach to define R_{equ} based on the conductance, G, of the pore or throat that is exact for a circular cylinder:

$$R_{equ} = \left(\frac{8G}{\pi}\right)^{1/4} \qquad (3)$$

In a network of pores and throats we do not know each pressure drop ΔP a priori. Hence to compute the flow and effective viscosities requires an iterative approach, developed by Sorbie et al. in their network model studies of non-Newtonian flow (Sorbie et al., 1989). An initial guess is made for the effective viscosity in each network element. The choice of this initial value is rather arbitrary but does influence the rate of convergence, although not the final results. As a general rule, when one is interested in solving for only one flow rate across the network, the initial viscosity guess can be taken as the limiting boundary condition, μ_0 (i.e. the viscosity at very low shear rates). However, when trying to explore results for a range of increasing flow rates, the convergence process can be significantly speeded up by retaining the last solved solution for viscosity.

Once each pore and throat has been assigned an effective viscosity and conductance, the relationship between pressure drop and flow rate across each element can be found.

$$Q_i = \frac{G_i}{\mu_{eff}^i}\Delta P_i \qquad (4)$$

By invoking conservation of volume in each pore with appropriate inlet and outlet boundary conditions (constant pressure), the pressure field is solved across the entire network using standard techniques. As a result the pressure drop in each network element is now known, assuming the initial guess for viscosity. Then the effective viscosity of each pore and throat is

updated and the pressure recomputed. The method is repeated until satisfactory convergence is achieved. In our case, convergence must be achieved
simultaneously in all the network elements. The pressure is recomputed if
the flow rate in any pore or throat changes by more than 1% between iterations. The total flow rate across the network Q_t is then computed and an
apparent viscosity is defined as follows:

$$\mu_{app} = \mu_N \frac{Q_N}{Q_t} \tag{5}$$

where Q_N is the total flow rate for a simulation with the same pressure
drop with a fixed Newtonian viscosity μ_N. The Darcy velocity is obtained
from $q = Q_t/A$, where A is the cross sectional area of the network.

3.1. NON-NEWTONIAN RESULTS

We predict the porous medium rheology of four different experiments in
the literature where the bulk shear-thinning properties of the polymers used
were also provided. Two of the experiments (Hejri et al., 1988; Vogel and
Pusch, 1981) were performed on sand packs and for these we used the sand
pack network and two were performed on Berea sandstone (Cannella et al.,
1988; Fletcher et al., 1991), for which the Berea network was used. Table II
lists the properties used to match the measured bulk rheology to a truncated power-law.

We can account for the permeability difference between our model and
the systems we wish to study by realizing that simply re-scaling the network
size will result in a porous medium of identical topological structure, but
different permeability. To predict the experiments we generated new networks with all lengths scaled by a factor $\sqrt{K^{exp}/K^{net}}$, where the superscripts *exp* and *net* stand for experimental and network, respectively. By
construction the re-scaled network now has the same permeability as the
experimental system, but otherwise has the same structure as before. Note
that this is not an *ad-hoc* procedure since the scaling factor is based on the
experimentally measured permeability.

Table II. Truncated power law parameters used to fit the experimental data

Experiment	C	n	μ_0 (Pa.s)	μ_∞ (Pa.s)	ϕ	K (D)
Hejri et al. (1988)	0.181	0.418	0.5	0.0015	0.34	0.525
Vogel and Pusch (1981)	0.04	0.57	0.1	0.0015	0.5	5
Fletcher et al. (1991)	0.011	0.73	0.012	0.0015	0.2	0.261
Cannella et al. (1988)	0.195	0.48	0.102	0.0015	0.2	0.264

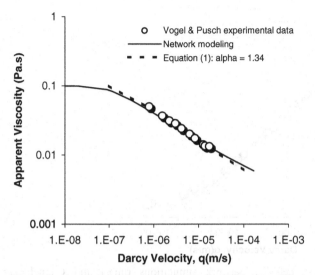

Figure 2. Comparison between network simulations (line) and the Vogel and Pusch (1981) experiments on a sand pack (circles). The dashed line is an empirical fit to the data, Equation (1), using an adjustable scaling factor α.

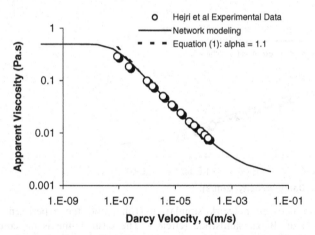

Figure 3. Comparison between network simulations (line) and the Hejri *et al.* (1988) experiments on a sand pack (circles). The dashed line is an empirical fit to the data, Equation (1), using an adjustable scaling factor α.

Figures 2–5 compare the predicted and measured porous medium rheology. Also shown are best fits to the data using Equation (1). Note that the empirical approach requires a medium-dependent parameter α to be defined, and does not accurately reproduce the whole shape of the curve. In one of the sandstone experiments – Figure 4 – the viscosity at low flow rates exceeds that measured in the bulk. This could be due to pore blocking

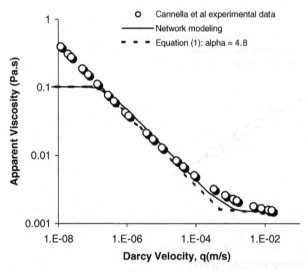

Figure 4. Comparison between network simulations (lines) and Cannella *et al.*
(1988) experiments on Berea sandstone (circles). The dashed line is an empirical fit
to the data, Equation (1), using an adjustable scaling factor.

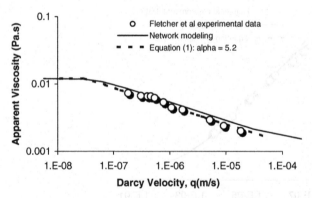

Figure 5. Comparison between network simulations (line) and the experiments of
Fletcher *et al.* (1991) on Berea sandstone (circles). The dashed line is an empirical fit to the data, Equation (1), using an adjustable scaling factor α.

by polymer adsorption that we do not model. We also slightly over-predict
the viscosity in the other Berea sample – Figure 5. Overall the predictions
– made with no adjustable parameters – are satisfactory and indicate that
the network model is capturing both the geometry of the porous medium
and the single-phase non-Newtonian rheology. In the next section we will
extend this approach to the more challenging case of two-phase flow, albeit
with Newtonian fluids.

4. Two-Phase Flow

Two and three-phase relative permeabilities for water-wet Berea sandstone have been measured by Oak (1990). In previous work we have shown that we can predict oil/water drainage and water flood relative permeabilities accurately (Blunt *et al.*, 2002; Valvatne and Blunt, 2004). In this case we know we have an appropriate network with a well-characterized wettability. The only issue is that during water injection a distribution of advancing oil/water contact angles has to be assumed – we find the uniform distribution of contact angles that matches the observed residual non-wetting phase saturation and from that predict both oil and water relative permeabilities. In this section we will show how to adjust the pore and throat size distributions to match two-phase data capillary pressure data and then predict relative permeability when we do not have an exact network representation of the medium of interest. In the following section we will predict three-phase data from Oak (1990).

When using pore-scale modeling to predict experimental data it is clearly important that the underlying network is representative of the rock. However, if the exact rock type has to be used for the network construction, the application of predictive pore-scale modeling will be severely limited due to the complexity and cost of methods such as X-ray microtomography. In this section we will use the topological information of the Berea network (relative pore locations and connection numbers) to predict the flow properties of a sand pack measured by Dury (1997) and Dury *et al.* (1998). We do not use our sand pack network, since in this case the network and the sand used in the experiments have very different properties. Capillary pressure data is used to tune the properties of the individual network elements.

Dury *et al.* (1998) measured secondary drainage and tertiary imbibition capillary pressure (main flooding cycles) and the corresponding non-wetting phase (air) relative permeabilities for an air/water system. The capillary pressures are shown in Figure 6 (Dury *et al.*, 1998). To predict the data, first all the lengths in the Berea network are scaled using the same permeability factor that was used for non-Newtonian flow. From Figure 6 it is, however, clear that the predicted capillary pressure is not close to the experimental data. This indicates the difficulty of predicting multiphase measurements – the capillary pressure and relative permeabilities are influenced by the distribution of pore and throat sizes as well as the absolute permeability. The distribution of throat sizes is subsequently modified iteratively until an adequate pressure match is obtained against the experimental drainage data (Figure 7), with individual network elements assigned inscribed radii from the target distribution while still preserving their rank order – that is the largest throat in the network is given the largest radius from the target distribution and so on. This should ensure that

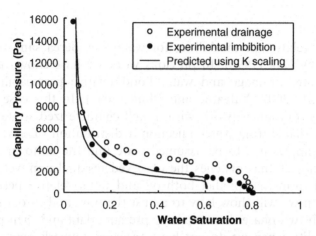

Figure 6. Comparison between predicted capillary pressures and experimental data by Dury *et al.* (1998). The size of the elements in the Berea network is modified using a scaling factor based on absolute permeability and the predictions are poor, indicating that the pore size distribution needs to be adjusted to match the data.

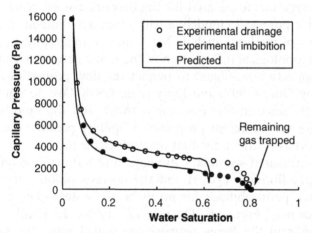

Figure 7. Comparison between predicted and measured (Dury *et al.*, 1998) capillary pressures following a network modification process to match the drainage data. Now the match is excellent, except at high water saturations. The trapped gas (air) saturation is 1 minus the water saturation when the capillary pressure is zero.

size correlations between individual elements and on larger scales are maintained. Modifications to the throat size distribution at each iteration step were done by hand rather than by any optimization technique. The results are insensitive to the details of how the throat sizes are adjusted – the final throat size distribution obtained was effectively a unique match since the rank order of size and connectivity was preserved.

Capillary pressure hysteresis is a function of both the contrast between pore body and throat radii and the contact angle hysteresis. We distribute advancing contact angles uniformly between 16 and 36 degrees, consistent with measured values by Dury (Dury, 1997; Dury et al., 1998), while keeping receding values close to zero. The radii of the pore bodies is determined from Valvatne and Blunt (2004)

$$r_p = \max \left(\beta \frac{\sum_{i=1}^{n_c} r_i}{n_c}, \max(r_i) \right), \tag{6}$$

where n_c is the connection number and β is the aspect ratio between the pore body radius r_p and connecting throat radii r_i. A good match to experimental imbibition capillary pressure is achieved by distributing the aspect ratios between 1.0 and 5.0 with a mean of 2.0. This distribution is very similar to that of the original Berea network, though with a lower maximum value, which in the original network was close to 50. This is expected as the Berea network has a much larger variation in pore sizes. The absolute size of the model, defining individual pore and throat lengths, is adjusted such that the average ratio of throat length to radius is maintained from the original network. Pore and throat volumes are adjusted such that the target porosity is achieved, again maintaining the rank order.

In Figure 8 the predicted air relative permeability for secondary drainage and tertiary imbibition are compared to experimental data by Dury et al. (1998). The experimental data were obtained by the stationary liquid method where the water does not flow, while air is pumped through the system and the pressure drop is measured. The relative permeability hysteresis is well predicted. In imbibition snap-off disconnects the non-wetting phase leading to a lower relative permeability than in drainage. However, there are two features that we fail to match. First, the experimental trapped air saturation is much lower than predicted by the network model (Figure 7) and is lower than the value implied by the extinction point in Figure 8. Second, the extinction and emergence (when air first starts to flow) saturations are different in the experiment, while the network model predicts similar values (Figure 8). This behavior is difficult to explain physically, as the network model predicts that the trapped air saturation and the emergence and extinction points are all consistent with each other. Dury (1997) suggested that air compressibility could allow trapped air ganglia to shrink as water is injected, leading to a small apparent trapped saturation. Furthermore, air could have escaped from the end of the pack, even if the air did not span the system, leading to displacement even when the apparent air relative permeability was zero. For lower water saturations where there is more experimental confidence in the

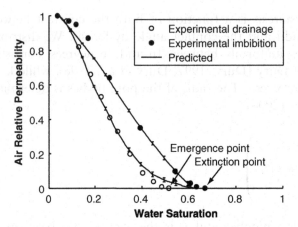

Figure 8. Comparison of network model air relative permeability predictions to experimental data measured on a sand pack by Dury *et al.* (1998). The flooding cycles shown are secondary drainage and tertiary imbibition (main cycles) and the experimental data are obtained using the stationary liquid method. The emergence point represents when gas first starts to flow during gas invasion (drainage) and the extinction point is where gas ceases to flow during imbibition.

data, the predictions are excellent and give confidence to the ability of pore-scale modeling to use readily available data (in this case capillary pressure) to predict more difficult to measure properties, such as relative permeability.

5. Three-Phase Flow

Three-phase – oil, water and gas – flow can be simulated in the network model (Piri and Blunt, 2002). All the different possible configurations of oil, water and gas in a single corner of a pore or throat are evaluated – Figures 9 and 10. Displacement is a sequence of configuration changes. For each change a threshold capillary pressure is computed (Piri and Blunt, 2002). The next configuration change is the one that occurs at the lowest invasion pressure of the injected phase. By changing what phase is injected into the network any type of displacement can be simulated (Piri and Blunt, 2002).

In this section we will predict steady-date three-phase relative permeability measured on Berea cores by Oak (1990). The two-phase oil/water data has already been predicted (Blunt *et al.*, 2002; Valvatne and Blunt, 2004) – we did not adjust any of the geometrical properties of the network (pore and throat sizes or shapes) and assumed that the receding oil/water contact angle was zero. As mentioned before, the distribution of advancing contact angles was adjusted to match the measured residual oil saturation.

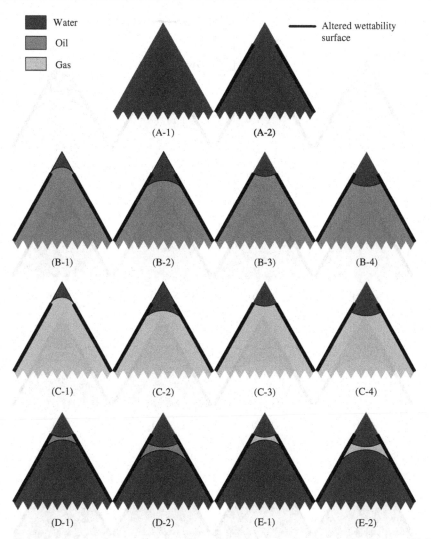

Figure 9. One- and two-phase configurations for a single corner. The bold solid line indicates regions of the surface with altered wettability. A phase may be present in the center of the pore space or as a spreading or wetting layer, sandwiched between other phases. Water is always present in the corner. The network model simulates a sequence of displacement events that represent the change from one configuration to another.

Piri and Blunt (2002) and Lerdahl *et al.* (2000) have presented three-phase predictions for this dataset – in this work we will consider a typical yet difficult case – gas injection after waterflooding – and compare predictions and experiment on a point-by-point basis. We assume that we have a spreading oil (Piri and Blunt, 2002) with oil/water and gas/oil interfacial tensions typical of light alkane/water/air systems as studied by Oak (1990). Figure 11 shows

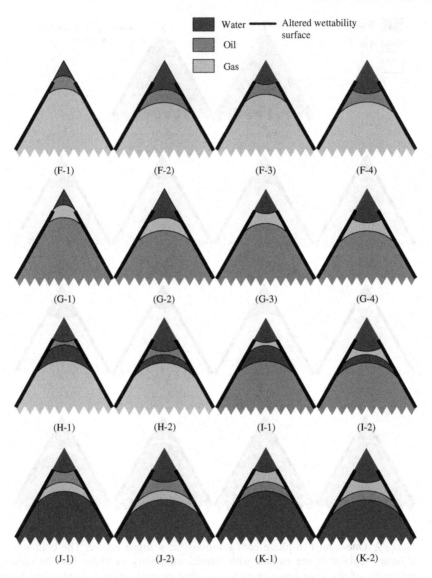

Figure 10. Three-phase configurations continued from Figure 9.

the saturation path for an experiment where gas is injected into waterflood residual oil. This is a particular challenge for pore-scale modeling since at the beginning of the displacement some of the oil is trapped and must become reconnected at the pore scale through double drainage and oil layer formation before it can be displaced (Lerdahl *et al.*, 2000; Piri and Blunt, 2002; van Dijke *et al.*, 2004). The network model simulates gas displacing either water or oil in order to track the saturation path seen experimentally.

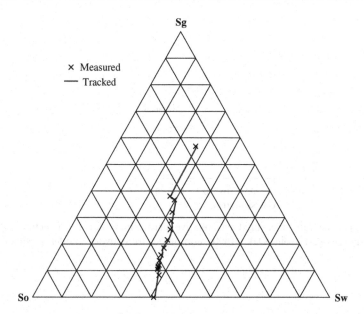

Figure 11. Saturation path for a steady-state experiment by Oak (1990) for gas injection into oil and water (crosses). The network model undergoes a series of displacements of water or oil by gas to reproduce a similar path (line).

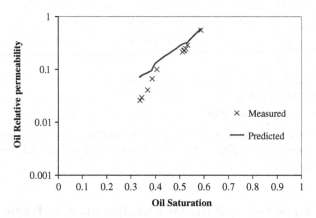

Figure 12. Experimentally measured oil relative permeability for gas injection after waterflooding (crosses) from Oak (1990) compared to predictions from network modeling (line).

Figures 12–14 show the predicted and measured three-phase oil, gas and water relative permeabilities respectively. The quality of the predictions is similar to that obtained for gas injection into higher initial oil saturations (Piri and Blunt, 2002). The three-phase predictions are satisfactory, although not as good as for two-phase flow. This is because the pore scale physics is more complex and uncertain when three phases are flowing – in

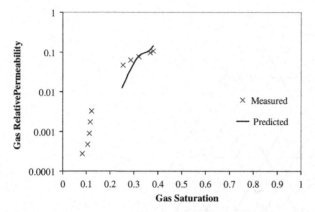

Figure 13. Experimentally measured gas relative permeability for gas injection after waterflooding (crosses) from Oak (1990) compared to network model predictions (line).

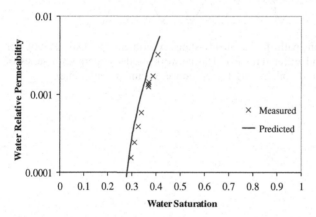

Figure 14. Experimentally measured water relative permeability for gas injection after waterflooding (crosses) from Oak (1990) compared to network model predictions (line).

particular we do not know how well the fluid configurations in Figure 10 represent the true arrangements of fluid.

For the oil relative permeability, Figure 12, the network model tends to over-predict the relative permeability at low saturation. At low oil saturation the oil is flowing in layers (see Figure 10G) and the relative permeability is controlled by our assumptions of layer connectivity and conductance. It appears that we over-estimate the connectivity of the layers and that in reality oil layers do not have the rather high effective conductance that we assume using our idealized model of the pore space geometry. The gas relative permeability is well predicted at high gas saturation. However, at low saturation finite size effects in the network mean that we predict that gas is not connected even though it does flow in the experiments. The predictions of water relative permeability are excellent.

6. Discussion and Conclusions

We have demonstrated that pore-scale modeling combined with geologically realistic networks can reliably predict single-, two- and three-phase data for water-wet media. The predictions of the single-phase shear thinning rheology of polymer solutions in porous media were excellent and superior to empirical scaling approaches, since the results did not depend on an unknown porous-medium-dependent factor. For two-phase flow the results were also excellent if the pore structure of the porous medium is known, although the distribution of contact angles for waterflooding needs to be estimated. If the pore structure is not known *a priori*, we showed how to adjust the pore size distribution to match capillary pressure data and then use this to make good predictions of relative permeability. For three-phase flow modeling is more of a challenge because of the complexity of the pore-scale physics. However, we were able to predict relative permeabilities with reasonable accuracy for gas injection after waterflooding.

We did not address wettability in this paper. Most natural media that have been in contact with oil or other non-aqueous phase liquids change their wettability and often display mixed-wet or oil-wet characteristics. The network model presented in this paper can handle media of any wettability and has made accurate predictions of relative permeability and oil recovery for mixed-wet reservoir samples (Piri and Blunt, 2002; Valvatne and Blunt, 2004). Figures 9 and 10 show all the possible two- and three-phase configurations including wettability alteration: after primary drainage, where oil contacts the solid surface directly the oil/water contact angle may change to any value (indicated by the bold lines in Figures 9 and 10). During gas injection different gas/oil and gas/water contact angles can also be assigned to these regions (Piri and Blunt, 2002). Regions of the pore space that remain water filled remain water-wet. Extensive experimental verification of the models for media of different wettability, particularly for three-phase flow, remains the subject of future study.

The aim of pore-scale modeling is to predict properties that are difficult to measure, such as relative permeability, from more readily available data, such as drainage capillary pressure. In addition, the model can readily be used to predict the changes in flow properties as the pore structure or wettability varies. As such it can be used to characterize multiphase properties in geological models. We have already shown that using pore-scale modeling to characterize variations in relative permeability leads to significantly different predictions of recovery at the field scale than traditional empirical modeling approaches (Blunt *et al.*, 2002).

Acknowledgements

The members of the Imperial College Consortium on Pore-Scale Modelling (BHP, Gaz de France, JNOC, PDVSA-Intevep, Schlumberger, Shell, Statoil, the UK Department of Trade and Industry and the EPSRC) are thanked for their financial support. We also thank Pål-Eric Øren (Statoil) for sharing his network data with us.

References

Adler, P. M. and Thovert, J. F.: 1998, Real porous media: Local geometry and macroscopic properties, *Appl. Mechanics Rev.* **51**(9), 537–585.

Bakke, S. and Øren, P. E.: 1997, 3-D pore-scale modelling of sandstones and flow simulations in the pore networks, *SPE Journal* **2**(2), 136–149.

Blunt, M. J.: 1998, Physically-based network modeling of multiphase flow in intermediate-wet porous media, *J. Petrol. Sci. Eng.* **20**(3–4), 117–125.

Blunt, M. J., Jackson, M. D., Piri, M. and Valvatne, P. H.: 2002, Detailed physics, predictive capabilities and macroscopic consequences for pore-network models of multiphase flow, *Adv. Water Resour.* **25**(8–12), 1069–1089.

Bryant, S. and Blunt, M.: 1992, Prediction of relative permeability in simple porous-media, *Phys. Rev. A* **46**(4), 2004–2011.

Bryant, S. L., King, P. R., and Mellor, D. W.: 1993a, Network model evaluation of permeability and spatial correlation in a real random sphere packing, *Transport in Porous Media* **11**(1), 53–70.

Bryant, S. L., Mellor, D. W., and Cade, C. A.: 1993b Physically representative network models of transport in porous-media, *AIChE Journal* **39**(3), 387–396.

Cannella, W. J., Huh, C. and Seright, S.: 1988, Prediction of xanthan rheology in porous media, Paper SPE 18089, in: *Proceedings of the 63rd Annual Technical Conference and Exhibition of the Society of Petroleum Engineers*, Houston, TX, USA.

Chatzis, I. and Dullien, F. A. L.: 1997, Modelling pore structures by 2-D and 3-D networks with application to sandstones, *J. Can. Petrol. Technol.* **16**(1), 97–108.

Delerue, J. F. and Perrier, E.: 2002, DXSoil, a library for 3D image analysis in soil science, *Comp. Geosci.* **28**(9), 1041–1050.

Dixit, A. B., McDougall, S. R., and Sorbie, K. S.: 1997, Pore-level investigation of relative permeability hysteresis in water-wet systems, SPE 37233, in: *Proceedings of the 1997 SPE International Symposium on Oilfield Chemistry*, February 1997, Houston.

Dixit, A. B. et al.: Pore-scale modeling of wettability effects and their influence on oil recovery, *SPE Reserv. Eval. Eng.* **2**(1), 25–36.

Dunsmuir, J. H. et al.: 1991, X-ray microtomography. A new tool for the characterization of porous media, SPE 22860, in: *Proceedings of the 1991 SPE Annual Technical Conference and Exhibition*, October 1991, Dallas.

Dury, O.: 1997, Organic pollutants in unsaturated soils: Effect of butanol as model contaminant on phase saturation and flow characteristics of a quartz sand packing, PhD Thesis, ETH Zürich.

Dury, O., Fischer, U. and Schulin, R.: 1998, Dependence of hydraulic and pneumatic characteristics of soils on a dissolved organic compound, *J. Contam. Hydrol* **33**(1–2), 39–57

Fischer, U. and Celia, M. A.: 1999, Prediction of relative and absolute permeabilities for gas and water from soil water retention curves using a pore-scale network model, *Water Resour. Res.* **35**(4), 1089–1100.

Fletcher, A. J. P. et al.: 1991, Measurements of polysaccharide polymer properties in porous media, Paper SPE 21018, in: *Proceedings of the Society of Petrol. Engineers International Symposium on Oilfield Chemistry*, Anaheim, California, USA.

Hejri, S., Willhite, G. P. and Green, D. W.: 1988, Development of correlations to predict biopolymer mobility in porous media, Paper SPE 17396, in: *Proceedings of the Society of Petroleum Engineers Enhanced Oil Recovery Symposium* , Tulsa, USA.

Jerauld, G. R. and Salter, S. J.: 1990, Effect of pore-structure on hysteresis in relative permeability and capillary pressure. Pore-level modeling, *Transport in Porous Media* **5**(2), 103–151.

Lerdahl, T. R., Øren, P. E., and Bakke, S.: 2000, A predictive network model for three-phase flow in porous media, SPE 59311, in: *Proceedings of the SPE/DOE Symposium on Improved Oil Recovery*, April 2–5, Tulsa.

Lopez, X. P., Valvatne, P. H. and Blunt, M. J.: 2003, Predictive network modeling of single-phase non-Newtonian flow in porous media, *J. Colloid and Interf. Sci.* **264**(1), 256–265.

Oak, M. J.: 1990, Three-phase relative permeability of water-wet Berea, SPE 20183, in: *Proceedings of the SPE/DOE Seventh Symposium on Enhanced Oil Recovery*, 28 April, Tulsa.

Øren, P. E. and Bakke, S.: 2002, Process based reconstruction of sandstones and prediction of transport properties, *Transport in Porous Media* **46**(2–3), 311–343.

Øren, P. E., Bakke, S. and Arntzen, O. J.: 1998, Extending predictive capabilities to network models, *SPE Journal* **3**(4), 324–336.

Patzek, T. W.: 2001, Verification of a complete pore network simulator of drainage and imbibition, *SPE Journal* **6**(2), 144–156.

Pearson, J. R. A. and Tardy, P. M. J.: 2002, Models for flow of non-Newtonian and complex fluids through porous media, *J. Non-Newton. Fluid Mechanics* **102**: 447–473.

Piri, M. and Blunt, M. J.: 2002, Pore-scale modeling of three-phase flow in mixed-wet systems, Paper SPE 77726, in: *Proceedings of the SPE Annual Meeting*, 29 September–2 October San Antonio, Texas.

Sorbie, K. S.: 1991, *Polymer-Improved Oil Recovery,* CRC Press Inc.

Sorbie, K. S., Clifford, P. J. and Jones, E. R. W.: 1989, The rheology of pseudoplastic fluids in porous media using network modeling, *J. Colloid Interf. Sci.* **130**(2): 508–534.

Spanne, P. et al.: 1994, Synchrotron computed microtomography of porous-media – topology and transports, *Phy. Rev. Lett.* **73**(14), 2001–2004.

Valvatne, P. H. and Blunt, M. J.: 2004, Predictive pore-scale network modeling of two-phase flow in mixed wet media, *Water Resources Res.* **40**, W07406, DOI: 10.1029/2003 WR002627.

van Dijke, M. I. J., Sorbie, K. S., Sohrabi, M. and Danesh, A.: 2004, Three-phase flow in WAG processes in mixed-wet porous media: pore scale network simulations and comparison with water-wet micromodel experiments, SPE 75192, in: *Proceedings of the SPE/DOE Conference on Improved Oil Recovery*, April, Tulsa, OK.

Vogel, H. J.: 2000, A numerical experiment on pore size, pore connectivity, water retention, permeability, and solute transport using network models, *Eur. J. Soil Sci.* **51**(1), 99–105.

Vogel, P. and Pusch, G.: 1981, Some aspects of non-Newtonian fluids in porous media, in: *Proceedings of the First European Symposium on EOR*, September 1981, Bournemouth, UK.

Wilkinson, D. and Willemsen, J. F.: 1983, Invasion percolation – a new form of percolation theory, *J. Phy. A: Math. Gen.* **16**(14), 3365–3376.

Transp Porous Med (2005) 58:43–62
DOI: 10.1007/s11242-004-5469-1

Digitally Reconstructed Porous Media: Transport and Sorption Properties

M. E. KAINOURGIAKIS[1,*], E. S. KIKKINIDES[2], A. GALANI[1,3],
G. C. CHARALAMBOPOULOU[1] and A. K. STUBOS[1]
[1]National Center for Scientific Research "Demokritos", Institute of Nuclear Technology
and Radiation Protection, 15310 Ag. Paraskevi Attikis, Athens, Greece
[2]Chemical Process Engineering Research Institute, Center for Research and Technology
Hellas, P.O. Box 361, Thermi-Thessaloniki 57001, Greece
[3]School of Chemical Engineering, National Technical University of Athens, 15780 Athens,
Greece

(Received: 13 October 2003; in final form 22 April 2004)

Abstract. The basic aim of this work is to present a combination of techniques for the reconstruction of the porous structure and the study of transport properties in porous media. The disordered structure of porous systems like random sphere packing, Vycor glass and North Sea chalk, is represented by three-dimensional binary images. The random sphere pack is generated by a standard ballistic deposition procedure, while the chalk and the Vycor matrices by a stochastic reconstruction technique. The transport properties (Knudsen diffusivity, molecular diffusivity and permeability) of the resulting 3-dimensional binary domains are investigated through computer simulations. Furthermore, physically sound spatial distributions of two phases filling the pore space are determined by the use of a simulated annealing algorithm. The wetting and the non-wetting phases are initially randomly distributed in the pore space and trial-and-error swaps are performed in order to attain the global minimum of the total interfacial energy. The effective diffusivities of the resulting domains are then computed and a parametric study with respect to the pore volume fraction occupied by each phase is performed. Reasonable agreement with available data is obtained in the single- and multi-phase transport cases.

Key words: reconstruction, diffusion, permeability, simulated annealing, adsorption

1. Introduction

The structural characterization and prediction of sorption and transport properties in porous materials is of great importance in various fields of technological and environmental interest such as catalysis, separations, oil recovery, aging of building materials, study of hazardous waste repositories etc. A large number of theoretical and experimental studies concerning

---

*Author for correspondence: e-mail: kainourg@chem.demokritos.gr

transport in porous materials and the dependence of the transport coefficients on the main structural parameters of the media have been published. The evolution of the modeling approaches for the representation of the porous structure, is a result of advances in theoretical and experimental techniques as well as in computational resources. A turning point in this respect has been the development of *statistical* methods that generate binary 3-dimensional images based on certain statistical properties of the medium, usually the porosity and the two-point correlation function (Alder *et al.*, 1990; Alder, 1992). The experimental information on the aforementioned statistical properties can be obtained either directly from SEM or TEM images or indirectly by small angle scattering techniques (SAS) (Levis *et al.*, 1991; Levitz and Tchoubar, 1992).

In contrast to the statistical methods, *process-based* models try to account for the physical processes underlying the formation of certain microstructure. Recent attempts include the generation of controlled porous glasses (CPG's) through a dynamic simulation of the actual spinodal decomposition process (Gelb and Gubbins, 1998), which is believed to be the main mechanism responsible for the formation of these materials. In the same context (although using a more phenomenological and thus less computationally intensive approach) is the grain consolidation model (Roberts and Schwartz *et al.*, 1985), which focuses on the modeling of diagenetic processes. These process-based reconstruction methods, although more sound from a physical point of view, suffer frequently from severe computational requirements and are limited to the specific material considered in each case.

In many cases of practical interest, however, the pore volume is filled by more than one fluid phases and the accurate representation of the porous structure is not the only parameter to be concerned with. The detailed distribution of the phases that occupy the pores strongly affects the overall transport properties and a reliable method must be applied for its determination. Recently, Lattice-Boltzmann two-phase flow simulations have been used to obtain the spatial distribution of the phases (Bekri *et al.*, 2002; Martys, 1999), however, their demands on computational resources are usually very high. For the simplest case, where only two phases are present in the pores, one wetting and one non-wetting, Knight *et al.* (1990), have proposed an attractive mesoscopic technique, assuming that the desired distribution is characterized by minimal total free interfacial energy. Silverstein and Fort (2000a,b) applied this method for the prediction of air-water interfacial area in wet unsaturated porous media, Mohanty (1997) calculated the thermal conductivity of 2-dimensional cuts of geologic media while Berkowitz and Hansen (2001) studied the distribution of water in partially saturated sandstone microstructures.

In this chapter the main principles (without focusing on details) of the aforementioned techniques are presented. Furthermore, their combination is employed to study in an efficient and sufficiently accurate manner transport properties in porous media.

2. Reconstruction of Porous Structure

2.1. REPRESENTATION OF POROUS STRUCTURE

The spatial distribution of matter in a porous medium can be typically represented by the phase function $Z(\mathbf{x})$, defined as follows:

$$Z(\mathbf{x}) = \begin{cases} 1 \text{ if } \mathbf{x} \text{ belongs to the pore space} \\ 0 \text{ otherwise} \end{cases} \tag{1}$$

where $\mathbf{x}$ is the position vector from an arbitrary origin.

Due to the disordered nature of porous media, $Z(\mathbf{x})$ can be considered as a stochastic process, characterized by its statistical properties. The porosity, ε, and the auto-correlation function $R_z(\mathbf{u})$ can be defined by the statistical averages (Alder, 1992):

$$\varepsilon = \langle Z(\mathbf{x}) \rangle \tag{2a}$$

$$R_z(\mathbf{u}) = \frac{\langle (Z(\mathbf{x}) - \varepsilon) \cdot (Z(\mathbf{x} + \mathbf{u}) - \varepsilon) \rangle}{\varepsilon - \varepsilon^2} \tag{2b}$$

Note that < > indicates spatial average. For an isotropic medium, $R_z(\mathbf{u})$ becomes one-dimensional as it is only a function of $u = |\mathbf{u}|$ (Alder, 1992).

2.2. STOCHASTIC RECONSTRUCTION

The purpose of the stochastic reconstruction procedure is the generation of a digitized 3-dimensional snapshot of $Z(\mathbf{x})$ with a specified statistical behavior which is assumed to be described by the first two moments of $Z(\mathbf{x})$, the mean value and the two point correlation function. The algorithm used for the reconstruction was first proposed by Joshi (1974) and was extended in three dimensions by Quiblier (1986) and Alder et al. (1990). The details of this algorithm can be found in the above references and only the main idea is outlined here. The space is discretized in N^3 cubic elements, the position of which is characterized by the vector $\mathbf{x}' = (i, j, k)$ where i, j, k integers with values $1, 2, \ldots, N$ and a random value $X(\mathbf{x}')$ is assigned to any element. The values $X(\mathbf{x}')$ are uncorrelated and normally distributed with a mean equal to 0 and a variance equal to 1. A correlated field Y with a correlation function $R_y(u)$ can be deduced from the X field by the inverse Fourier transform:

$$Y(\mathbf{x}') = N^{3/2} \sum_{\mathbf{m}} \left(\hat{R}_{y\mathbf{m}}\right)^{1/2} \cdot \hat{X}_{\mathbf{m}} \cdot e^{-2i\pi \mathbf{k}_{\mathbf{m}} \mathbf{x}'} \tag{3}$$

where $\hat{R}_{y\mathbf{m}}$ and $\hat{X}_{\mathbf{m}}$ are the coefficients of the discrete Fourier transform of R_y and X, respectively. The values $Y(\mathbf{x}')$ are real and normally distributed with zero mean and unit variance, hence the distribution function $P(y)$ is given by:

$$P(y) = (2\pi)^{-1/2} \int_{-\infty}^{y} e^{-t^2/2} \, dt \tag{4}$$

The extraction of the binary phase function $Z(\mathbf{x}')$ from the real array Y can be accomplished by the condition:

$$Z(\mathbf{x}') = \begin{cases} 1 & \text{if } P[Y(\mathbf{x}')] \leqslant \varepsilon \\ 0 & \text{otherwise} \end{cases} \tag{5}$$

The most difficult step of the overall technique is the determination of the correlation function $R_y(u)$ from the experimentally observed $R_z(u)$. After some tedious manipulations (Alder et $al.$, 1990) $R_z(u)$ can be expressed as a series of $R_y(u)$, specifically:

$$R_z(u) = \sum_{m=0}^{\infty} C_m^2 \cdot R_y^m(u) \tag{6}$$

The coefficients C_m are given by

$$C_m = (2\pi m!)^{-1/2} \int_{-\infty}^{+\infty} c(y) e^{-y^2/2} H_m(y) dy \tag{7}$$

where

$$c(y) = \begin{cases} \dfrac{\varepsilon - 1}{[\varepsilon(1-\varepsilon)]^{1/2}} & \text{if } P(y) \leqslant \varepsilon \\ \dfrac{\varepsilon}{[\varepsilon(1-\varepsilon)]^{1/2}} & \text{if } P(y) > \varepsilon \end{cases} \tag{8}$$

and $H_m(y)$ is the Hermite polynomial of mth order:

$$H_m(y) = (-1)^m e^{y^2/2} \frac{d^m}{dy^m} e^{-y^2/2} \tag{9}$$

Ideally, a representative reconstruction of a porous medium in three dimensions should have the same correlation properties as those measured on a single two-dimensional section, expressed properly by the various moments of the phase function. In practice, matching of the first-two moments, that is, porosity and auto-correlation function, has been customarily pursued since the first application of the method. This simplification however is not generally valid as one can find examples of porous media exhibiting quite different morphological properties while sharing the same $R_z(u)$ ((Kainourgiakis *et al.*, 2000)). In this case one should try to match multi-point correlation functions. Such an approach is however tedious rendering the whole exercise quite difficult to handle. Instead, the determination of the chord length distribution function, $p_i(r)$, which gives the probability for a chord of length r to lie in the phase i of the medium, has been pursued, by the calculation of the length of a sufficiently large number of random line segments that lie in the mass or pore phase respectively, with ends at the solid–void interface. Such a property is related to the multi-point correlation functions and can be easily determined in digitized biphasic media (Pavlovitch *et al.*, 1991; Levitz and Tchoubar, 1992; Coker and Torquato, 1995). In recent publications Yeong and Torquato (1998a, b) and Manwart *et al.* (2000) have incorporated the use of both the two point correlation function and the lineal-path function in the reconstruction process for a Fontainebleau sandstone sample.

In this work two porous materials are stochastically reconstructed, with typical pore sizes in the nano- (Vycor glass) and micrometer (North Sea chalk) regimes. As already mentioned above, the stochastic reconstruction procedure requires as input the porosity and the autocorrelation function of the actual porous material. For both media the autocorrelation function, $R_z(u)$ is taken from the literature (Kainourgiakis *et al.*, 1999; Bekri *et al.*, 2000). The porosity of the chalk is 0.383 while that of Vycor is 0.28. In Figures 1 and 2, images of the reconstructed 3-D domains are shown

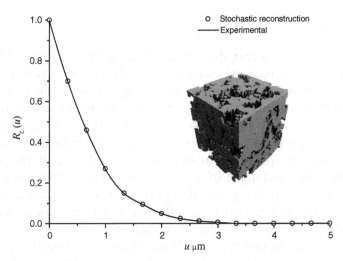

Figure 1. Image and two point autocorrelation function of reconstructed chalk with porosity 0.383. The pixel length is 0.333 μm.

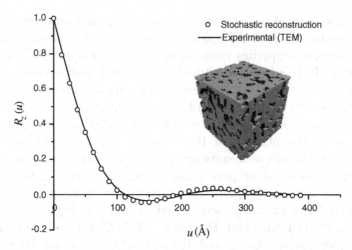

Figure 2. Image and two point autocorrelation function of reconstructed Vycor glass with porosity 0.28. The pixel length is 3 nm.

as well as the corresponding autocorrelation functions. The pixel size of the reconstructed domains is $0.333 \, \mu m$ and 3 nm for the chalk and Vycor respectively. In all cases, the agreement between the autocorrelation function of the reconstructed and the actual medium is excellent. Since the internal surface area, S_v, can be determined from the slope of $R_z(u)$ at zero distance, through the relation originally derived by Debye *et al.* (1957):

$$S_v = -4\varepsilon \, (1-\varepsilon) \, R_z'(0) \tag{10}$$

good agreement is obtained for this structural property too.

2.3. BALLISTIC DEPOSITION

Random packings of hard spheres, discs, and spheroids of prolate or oblate geometry have been the subject of considerable attention for many years, mainly due to their importance in powder technology (Vold, 1960; Visscher and Bolsterli, 1972) and in understanding the structure of liquids (Bernal, 1959) or amorphous, porous and random materials (Reiss, 1992; Bryant *et al.*, 1996). For the purposes of this work the random sequential deposition of non-overlapping spherical particles is employed as a simple process-based reconstruction technique of alumina membrane samples. In such a ballistic deposition, the packing rule differs from other methods in that the spheres position themselves under the influence of a unidirectional (vertical) force, rather than toward a center of attraction.

The basic idea of the algorithm is as follows: Balls are dropped sequentially from a random point well above the simulation box of size $L \times L \times L$. When a ball, i, is dropped, it hits ball m or the floor of the box in which case it stops. If it has contacted with ball m, it rolls down in a vertical plane on m until it is in contact with ball n. Then it rolls downwards in contact with both m and n until it hits ball p. If the contact with m, n and p is stable then ball i stops. If not, it rolls on the double contact that goes down most steeply, and so on. Such a procedure can be systematically followed through the use of a steepest-decent method followed by a conjugate gradient algorithm (Coelho et al., 1997). An alternative procedure has been pursued in the present study based on a Monte Carlo method: Each time we drop N "test" balls but allow only the one whose final position is lowest (provided that no overlapping with the spheres in the stack occurs) to remain and become a part of the stack. If N is large enough ($N > 10^5$) then we can recover random sphere packs with the same structural properties found by the more rigorous deposition algorithms. The number of spheres in the packs varied from 1000 up to 2000 and the length of the edge of the simulation box is ten times the sphere radius. Porosities ranging from 0.53 to 0.4 can be obtained with N ranging from 100 to 10^6. This Monte Carlo modification of the usual ballistic deposition algorithm offers significant computational time gains as well as programming simplicity.

In order to study structural properties such as correlation function and chord length distribution in the simulated random sphere packs, it is convenient to first digitize them and then work on the 3D digitized images. The simple algorithm proposed in Martys and Garboczi (1992), has been employed in the present study. Figure 3 shows the autocorrelation function and the respective 3D image of the medium.

The autocorrelation function, $R_z(u)$, obtained is in excellent agreement with the respective ones obtained elsewhere (Coelho et al., 1997) on a random sphere pack of $\varepsilon \sim 0.4$, generated by the more rigorous and time

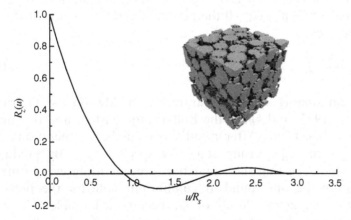

Figure 3. Digitized sphere pack domain and the corresponding two-point autocorrelation function. The porosity is equal to 0.41.

consuming ballistic deposition algorithm. Hence we observe first zero cross-ing in $R_z(u)$ around $u/R_s = 0.9$ followed by an anti-correlation up to $u/R_s = 2$. For $u/R_s > 3$, $R_z(u)$ drops practically to zero indicating the absence of significant correlation at distances larger than 1.5 particle diam-eters.

3. Spatial Distribution of the Pore Filling Phases

The distribution of the fluid phases in the reconstructed porous media can be achieved assuming that the total interfacial free energy, G_s, is minimal. Denot-ing the solid phase of the porous medium as σ, the wetting phase as α, the non-wetting phase as β and the interfacial free energy per unit area of $\sigma\alpha$, $\sigma\beta$, $\alpha\beta$ interfaces $\gamma_{\sigma\alpha}$, $\gamma_{\sigma\beta}$, $\gamma_{\alpha\beta}$ respectively, the following condition is fulfilled:

$$\gamma_{\sigma\alpha} = \gamma_{\sigma\beta} - \gamma_{\alpha\beta} \tag{11}$$

The function G_s then, can be evaluated by:

$$G_s = \sum_n A_n \gamma_n \tag{12}$$

where A_n is the elementary interfacial area and $n = \sigma\alpha$, $\sigma\beta$ or $\alpha\beta$.
To determine the minimum of G_s the method of simulated annealing (SA), introduced by Kirkpatrick et al. (1983) for the minimization of multidimen-sional functions, is employed. Initially, a specified number of voxels, in ran-dom positions of the pore space, are marked as wetting sites, the rest ones are marked as non-wetting. The number of wetting and non-wetting sites corresponds to a desired volume fraction occupied by each phase (denoted as S_α and S_β for α- and β- phases, respectively). Then a wetting and a non-wetting site exchange their positions. This change results in a variation of G_s by ΔG_s. If at a certain step $\Delta G_s \leqslant 0$, the new configuration is uncon-ditionally accepted while if $\Delta G_s > 0$ the new configuration is accepted with a probability given by:

$$P = \exp\left(-\frac{\Delta G_s}{G_{ref}}\right) \tag{13}$$

where G_{ref} is an analog of $k_B T$ parameter in Metropolis algorithm (Metropolis et al., 1953) and k_B, T the Boltzmann constant and the ambi-ent temperature, respectively. After a sufficient number of iterations, the system approaches the equilibrium state, for specific G_{ref}. By gradually decreasing G_{ref} and repeating the simulation process, using every time as initial configuration the one found as equilibrium state for the previous G_{ref} value, new lower energy levels of G_s become achievable. The pro-cess is considered complete when despite the change in G_{ref}, the num-ber of accepted changes in different configurations becomes lower than a pre-specified value (the ratio of the number of acceptable moves to the total number of trials becomes lower than 10^{-5}).

The numerical scheme based on simulated annealing as described above, can be used for the determination of the condensate configuration during capillary condensation where α- phase corresponds to the condensate while β-corresponds to its vapor. Figure 4 illustrates the picture obtained from the simulation procedure for a single slit-like pore. Initially, a certain fraction of the void space is occupied by condensate distributed in a random mode (Figure 4(a)) and the system is evolving towards the minimum of total interfacial energy. The initial random distribution of the condensate clearly converges to the formation of a bulk liquid phase, separated from the gas phase by a concave meniscus (Figure 4(b)).

Recalling that during adsorption in mesoporous media capillary condensation occurs above a certain relative pressure, the simulated annealing can be used for the determination of the spatial distribution of adsorbate in reconstructed domains. Three-dimensional "wet" Vycor and alumina (sphere pack) domains, obtained during the course of the SA are shown in Figure 5. It is clear that when equilibrium is reached (starting from the random fluid configurations of Figures 5(a) and (c)) the adsorbate forms clusters localized in narrow pores (Figures 5(b) and (d)), in accordance to the basic principles of adsorption in mesoporous media. The adsorbate accumulation is further elucidated by Figures 5e and 5f, where the solid phase has been subtracted from the alumina images.

4. Determination of Transport Properties

4.1. FLOW IN DARCY'S REGIME

On the macroscopic level, the superficial velocity, $\mathbf{q}$, of a viscous fluid in a given sample of homogeneous and isotropic porous material is described by Darcy's law:

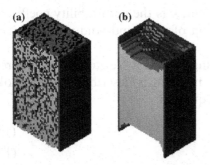

Figure 4. 3D simulated images of a single partially filled rectangular pore through the course of SA simulation: (a) starting random configuration, (b) final configuration (for illustrational purposes only one half of the pore is shown).

Figure 5. Spatial distribution of a condensed adsorbate in 3-D reconstructed Vycor (a, b) and random sphere pack (c, d, e, f) images as produced by the SA simulation. The left column refers to the corresponding starting random configurations and the right to the final, optimized ones.

$$\mathbf{q} = -\frac{k}{\eta} \cdot \nabla p \tag{14}$$

where ∇p is a prescribed pressure gradient, k is the permeability coefficient, which depends on the spatial distribution of solid and void phase and η is the fluid viscosity.

Turning to the microscale, the creeping flow of a Newtonian incompressible fluid within the pore space of the medium is described by the Stokes equation coupled with the continuity equation:

$$\eta \nabla^2 \mathbf{v} = \nabla p \tag{15a}$$
$$\nabla \cdot \mathbf{v} = 0 \tag{15b}$$

where $\mathbf{v}$ and p are the local fluid velocity vector and the pressure, respectively. The boundary conditions for $\mathbf{v}$ are spatial periodicity and no-slip at the surface of the solid unit elements. Hence, in order to determine the permeability from Equation (14), one needs to determine first the superficial velocity by calculating the flow field from Equations (15) with the appropriate boundary conditions.

The numerical solution of Equations (15) is achieved by the use of a finite-difference scheme in conjunction with the artificial compressibility relaxation algorithm (Alder et al., 1990). The pore space is discretized through a marker-and-cell (MAC) mesh with the pressure defined at the center of the cell, and the velocity components defined along the corresponding face boundaries. The resulting linear system of equations is solved by the successive overrelaxation method (Press et al., 1992).

In Table I the fluid permeability values are presented. The calculated permeability of the stochastichally reconstructed domains of Vycor lies within the experimentally determined range reported by Lin et al. (1992) while for the ballistic deposition structures (alumina) an excellent agreement is observed with the predictions of the well known Blake-Kozeny equation (Bird et al., 1960):

$$k = \frac{4R_S^2}{180} \cdot \frac{\varepsilon^3}{(1-\varepsilon)^2} \qquad (16)$$

where, R_S is the radius of each sphere of the random sphere pack.

4.2. DIFFUSION IN THE KNUDSEN AND MOLECULAR REGIMES

The effective diffusivities of inert tracers in the reconstructed porous structures are determined from the mean-square displacement $\langle \mathbf{r}^2 \rangle$, of a statistically sufficient number of identical point-like molecules injected in the void space of the medium, according to the well known equation:

$$D = \lim_{t \to \infty} \frac{\langle \mathbf{r}^2 \rangle}{6t} \qquad (17)$$

Table I. Calculated and reported permeabilities for the γ-alumina membrane ($\varepsilon = 0.42$), for the porous Vycor glass ($\varepsilon = 0.28$) and for the North Sea Chalk ($\varepsilon = 0.383$)

	$k(m^2)$	
	Simulation	Reported values
Alumina membrane	4.3×10^{-19}	4.9×10^{-19} (from Equation (16) for $R_S = 10$ nm)
Vycor	7.8×10^{-20}	$4 \times 10^{-20} - 1.0 \times 10^{-19}$ (Lin et al., 1992)
North Sea Chalk	5.5×10^{-15}	5.68×10^{-15} (Bekri et al., 2000)

where t is the travel time of the molecules. For the case of Knudsen gas diffusion this quantity is monitored through the distance traveled by the molecules assuming that they move at a constant speed equal to the thermal speed $u_0 = (8k_B T/\pi m)^{1/2}$, as indicated in similar studies (Tomadakis and Sotirchos, 1993; Kainourgiakis et al., 1999, Kainourgiakis et al., 2000), where k_B is the Boltzmann constant, T the ambient temperature and m the molecular mass. The travel time is related to the mean thermal speed by the equation $t = s/u_0$ hence Equation (17) can be written as:

$$D = \lim_{s \to \infty} \frac{\langle \mathbf{r}^2 \rangle}{6s} u_0 \qquad (18)$$

A random position in the pore space is first defined as the initial position of the molecule to travel within the porous medium. Subsequently, direction angles are randomly assigned to the molecule which starts its random walk moving from voxel to voxel along this direction. At each step a check is made whether the molecule hits a solid wall and if this happens it undergoes a diffuse reflection according to the cosine law (Bird, 1976).

For the determination of the molecular diffusivity of inert tracers the "blind random walk" method is applied (Schwartz et al., 1995). For a specific walker, a random pixel of the void phase is selected as initial position. At every time t a random trial move, of length l, to one of the six neighboring pixels is performed and the time is increased by δt which is selected to satisfy the condition $l^2 = 6D_0\delta t$, where D_0 the diffusivity of the tracer at the phase that occupies the void phase. The trial position is accepted unconditionally for moves from void to void pixels but for the case that the trial position lies in the solid phase the attempt move is not allowed. The procedure is repeated for sufficiently large number of time steps and the ratio D/D_0 is computed by:

$$\frac{D}{D_0} = \lim_{n \to \infty} \frac{\langle \mathbf{r}^2/l^2 \rangle}{n} \qquad (19)$$

where $n = t/\delta t$.

The travel time has to be large enough to ensure that the molecules feel the effect of all the structural details of the porous medium, in determining the macroscopic diffusivities. In this sense, the material can be considered as macroscopically homogeneous in terms of its structural and diffusion characteristics.

At all times a test is made to determine whether the molecule reaches the boundaries of the 3D medium where periodic boundary conditions have been employed.

A final point has to be made regarding the calculation of diffusivity. In general diffusivity calculations are based on the motion of the fluid in the pore space of the material. Nevertheless, in many cases effective diffusivity results have been reported in the literature which are basically diffusivities multiplied by the porosity of the material. In the remaining sections of this study we will use the term effective diffusivity, D_e, to include the porosity of the material, and distinguish it from the pure diffusivity values.

The calculated values of averaged effective diffusivity for each mesoporous reconstructed system are presented in Tables II and III. The experimental value of Knudsen diffusivity for the sphere pack model corresponds to the transport of He at 308 K through an alumina membrane (with porosity 0.42) fabricated by compaction of alumina microspheres of diameter 20 nm (Kainourgiakis *et al.*, 2000). For the Vycor glass membrane the porosity is equal to 0.28 and the experimental results used for comparison with the computed value correspond to diffusion of He at 298 K (Makri *et al.*, 1998).

The influence of porosity of the chalk structure on the diffusivity is presented in Figure 6. It is well known that the relation of diffusivity (or conductivity) with the porosity can be described by the empirical equation, originally proposed by Archie (1942).

$$\frac{D}{D_0} = A\varepsilon^m \tag{20}$$

where A and m empirical parameters. The value of A is usually near to unity and the exponent m varies from 1 to 3 depending on the microgeometry of the porous medium under consideration (Bekri *et al.*, 2002; Glover,

Table II. Calculated and experimental Knudsen diffusivities of He in the γ-alumina membrane ($\varepsilon = 0.42$) and in porous Vycor glass ($\varepsilon = 0.28$)

	D_e (Knudsen) $\times 10^3$ (cm^2/s)	
	Simulation	Experiment
Alumina membrane	7.7	8.0 (Papadopoulos, 1993)
Vycor	0.84	0.85 (Makri *et al.*, 1998)

The temperatures for the diffusion in alumina and in Vycor are 308 K and 298 K respectively.

Table III. Calculated and reported molecular diffusivities for the γ-alumina membrane ($\varepsilon = 0.42$), for the porous Vycor glass ($\varepsilon = 0.28$) and for the North Sea Chalk ($\varepsilon = 0.383$)

	D_e (molecular)/D_0	
	Simulation	Values from literature
Alumina membrane	0.25	0.25 (Kim and Torquato, 1992)
Vycor	0.062	0.064 (Lin *et al.*, 1992)
North Sea Chalk	0.11	0.131 (Bekri *et al.*, 2000)

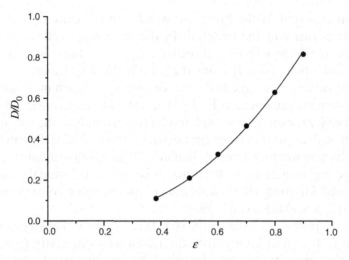

Figure 6. The diffusivity dependence on the porosity of reconstructed chalk matrix.

et al., 2000; Boving and Grathwohl, 2001) although some extreme values like 7.56 have been reported (Rosanne *et al.*, 2003). In the present study least square fitting of the results in Figure 6 yields $A = 1.063$ and $m = 2.35$. When two phases occupy the pore volume the relative diffusivity of an inert tracer for a given degree of saturation for either phase, S_k, where $k = \alpha$ or β, is determined through the relation:

$$D_R^k = \frac{D(S_k)}{D(S_k = 0)}$$

(21)

The simple (but very interesting) case where the tracer has zero diffusivity in one of the pore filling phases and finite to the other has been investigated. Figure 7 presents the corresponding relative molecular diffusivity of the sphere pack structure versus the saturation of the wetting phase. In the same figure computational results obtained by Martys (1999) for domains of non-overlapping spheres and experimental data for unconsolidated sand from the work of Leverett (1939) are also shown. It is observed that the diffusivity vanishes before the zero-diffusivity (non-conducting) phase totally occupies the pore space, indicating that below a given critical saturation the finite-diffusivity (conducting) phase forms isolated clusters not contributing to the transport. The critical saturation, $S_{k,c}$, of the conducting phase below which the overall diffusivity vanishes depends on the wettability of the conducting phase. When the conducting phase is wetting $S_{\alpha,c} \sim 0.15$ while when it is non-wetting $S_{\beta,c} \sim 0.3$. Additionally, Figure 7 shows that the relative diffusivity is more influenced by the degree of saturation when β-phase is the conducting one. The above lead to the conclusion that β-phase is less capable, compared to α-phase, to form sample-spanning clusters preferring to distribute itself in isolated blobs. The same

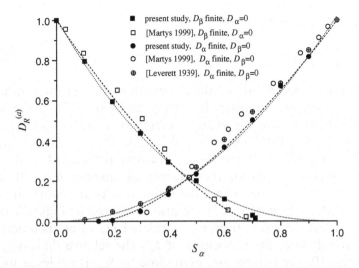

Figure 7. Relative diffusivity versus saturation of the α-phase for the sphere pack domain. Fitting with Equation (22) is presented by the dot line (...) while the dashed line (- - -) corresponds to fitting of Equation (23).

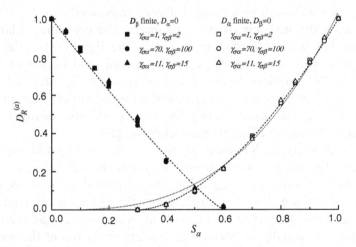

Figure 8. Relative diffusivity versus saturation of the α-phase for the stochastically reconstructed chalk. Fitting with Equation (22) is presented by the dot line (...) while the dashed line (- - -) corresponds to fitting of Equation (23).

behavior can be deduced for the reconstructed chalk (Figure 8) although the difference between the critical saturations is smaller in this case.

To correlate the relative diffusivity with the saturation of the finite-diffusivity phase the following empirical relations are used:

$$D_R^{(k)} = B S_k^n \tag{22}$$

where B, n empirical parameters and

$$D_R^{(k)} = \left(\frac{S_k - S_{k,c}}{1 - S_{k,c}} \right)^{\mu} \tag{23}$$

where μ empirical parameter and $S_{k,c}$ the critical saturation of the conducting phase below which D_R vanishes. By fitting Equation (22) to the calculated relative diffusivity when α-phase is conducting, we obtain for the
sphere pack domain $B = 1.022$ and $n = 2.082$ while for the reconstructed
chalk $B = 1.035$ and $n = 2.95$. When finite diffusivity is assigned to the β-
phase, the fitting process yields for the sphere pack domain $B = 1.01$ and
$n = 2.41$ but for the reconstructed chalk the fitting results are poor. The fitting functions based on Equation (22) are also presented in Figure 7 and
Figure 8, where a clear deviation from the calculated D_R values is observed
as $S_{k,c}$ is approached, since in the vicinity of $S_{k,c}$ the relative diffusivity is
expected to diverge. Better correlation, even close to $S_{k,c}$ is achieved using
Equation (23). For the random sphere pack the corresponding best-fitting
values are $\mu = 1.59$ when α-phase is conducting and $\mu = 1.41$ when diffusion occurs only in β-phase. Turning to reconstructed chalk we obtain $\mu =$
1.75 and $\mu = 1.1$ for diffusion in α- and β-phases respectively.

In Figure 8 an additional remark must be made. The computed relative
diffusivity values fall upon a single curve despite the significant variation
of γ_{ij} components. This demonstrates that the spatial distribution of the
phases is not affected significantly by the actual values of γ_{ij} involved (as
long as Equation (11) is satisfied), in agreement with the notion that under
equilibrium the fluid phases must occupy the same positions, regardless of
the absolute values of the individual interfacial energies.

The relative diffusivity in Knudsen regime can be investigated experimentally by partially filling a mesoporous matrix with a stationary condensed phase and subsequently measuring the permeability of a second
non-adsorbable gas, which does not condense in the pores, at least under
the specific conditions. This type of experiment is known as gas relative
permeability, and it is usually performed at low pressures where the mean
free path of the gas molecules is much greater than the characteristic pore
length. In this way, intermolecular collisions are rare and the transport process is determined primarily by the collisions between the molecules and
the solid walls. This mechanism resides within the Knudsen regime and
thus permeability reduces to Knudsen diffusivity.

Helium diffusivity was calculated for the case of binary domains representing the wet Vycor and alumina structures for various degrees of saturation, S_α.
The effect of the nature of the adsorbate on its spatial distribution within the
mesoporous matrix and therefore on the transport properties of the system,
was investigated by considering different sets of interfacial energies. For each
of the resulting systems the relative diffusivity, D_R (gas relative permeability)

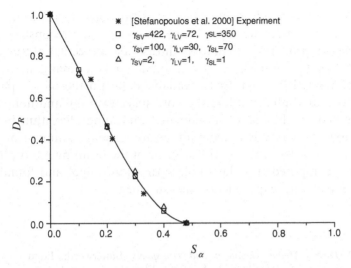

Figure 9. Simulated and experimental gas relative permeability curves of He for Vycor porous glass partially filled with a condensable adsorbate phase. The solid line is used only as a guide.

was determined by normalizing the effective He diffusivity computed for a certain S_α value by the one for the dry material ($S_\alpha = 0$). The simulation results are compared against experimental data obtained from the measurement of He permeability in Vycor preadsorbed with CH_2Br_2.

For the case of the reconstructed Vycor domains, a good agreement between simulation and experiment (Stefanopoulos *et al.*, 2000) occurs (Figure 9), demonstrating that the suggested approach can reproduce adequately the concentration profile of a condensable adsorbate in this specific type of material. The simulation results manage to capture not only the overall decay behaviour but also specific features of the experimental relative permeability curve (e.g. prediction of the percolation threshold, $S_{\alpha,c}$). In addition, it is interesting to point out again that despite the significant variation of γ_{ij} components, always under the condition of complete wetting ($\gamma_{\sigma\beta} > \gamma_{\sigma_\alpha}$), the computed relative permeability values fall upon a single curve.

5. Conclusions

A combination of techniques for the reconstruction of the porous structure and the study of transport properties in porous media is described. The disordered pore structure of alumina membranes (in the form of random sphere packs), Vycor porous glass and North Sea chalk, is represented by three-dimensional binary images generated on the basis of statistical information from the original material. The alumina membrane (random sphere pack) is generated by a standard ballistic deposition procedure, while the chalk and the Vycor matrices were produced by a stochastic reconstruction technique. Transport properties (Knudsen diffusivity, molecular diffusivity and permeability) of the resulting 3-dimensional binary domains are then computed and compared successfully to available data. Furthermore, spatial distributions of two fluid phases filling the pore space (wetting and non-wetting) are determined by the use of a simulated annealing algorithm based on the assumption of thermodynamic equilibrium (global minimum of the total interfacial energy). The effective diffusivities of the resulting multi-phase domains are then computed (for the whole saturation range) and found to be in good agreement with experimental measurements.

References

Adler, P. M.: 1992, *Porous Media: Geometry and Transports*. Butterworth, London.

Adler, P. M., Jacquin C. J. and Quiblier J. A.: 1990, Flow in simulated porous media. *Int. J. Multiphase Flow* **16**, 691.

Archie, G. E.: 1942, The electrical resistivity log as an aid in determining some reservoir characteristics. *Trans. AIME* **54**, 146.

Bekri, S., Vizika, O., Thovert, J.-F. and Adler P. M.: 2002, Binary two-phase flow with phase change in porous media, *Int. J. Multiphase Flow* **27**, 477.

Bekri, S., Xu, K., Yousefian, F., Adler, P. M., Thovert, J.-F., Muller, J., Iden, K., Psyllos, A., Stubos, A. K. and Ioannidis M. A.: 2000, Pore geometry and transport properties in North Sea chalk. *J. Petroleum Sci. Eng.* **25**, 107.

Berkowitz, B. and Hansen D. P.: 2001, A numerical study of the distribution of water in partially saturated porous rock. *Transport Porous Media* **45**, 301.

Bernal, J. D.: 1959, A geometrical approach to the structure of liquids. *Nature* **183**, 141.

Bird, G. A.: 1976, *Molecular Gas Dynamics*. Clarendon, Oxford.

Bird, R. B., Stewart, W. E. and Lightfoot E. N.: 1960, *Transport Phenomena*. John Wiley and Sons.

Boving, T. B. and Grathwohl, P.: 2001, Tracer diffusion coefficients in sedimentary rocks: correlation to porosity and hydraulic conductivity. *J. Contam. Hydrol.* **53**, 85.

Bryant, S, Mason, G and Mellor, D.: 1996, Quantification of spatial correlation in porous media and its effect on mercury porosimetry. *J. Colloid Interface Sci.* **177**, 88.

Coelho, D., Thovert, J.-F. and Adler, P. M.: 1997, Geometrical and transport properties of random packings of spheres and aspherical particles. *Phys. Rev. E* **55**, 1959.

Coker, D. A. and Torquato, S.: 1995, Simulation of diffusion and trapping in digitized porous media. *J. Appl. Phys.* **77**, 6087.

Debye, P., Anderson, H. R. and Brumberger, H.: 1957, *J. Appl. Phys.* **28**, 679.

Gelb, L. D., Gubbins, K. E.: 1998, Characterization of porous glasses: simulation models, adsorption isotherms, and the Brunauer–Emmett–Teller analysis method. *Langmuir* **14**, 2097.

Glover, P. W. J., Hole, M. J., Pous, J.: 2000, A modified Archie's law for two conducting phases. *Earth Planet Sci. Lett.* **180**, 369.

Joshi, M. Y.: 1974, A Class of stochastic models for porous media, PhD Thesis, University of Kansas.

Kainourgiakis, M. E., Kikkinides, E. S., Steriotis, Th.A., Stubos, A. K., Tzevelekos K. P. and Kanellopoulos, N. K.: 2000, Structural and transport properties of alumina porous membranes from process-based and statistical reconstruction techniques. *J. Colloid Interface Sci.* **231**, 158.

Kainourgiakis, M. E., Kikkinides, E. S., Stubos, A. K. and Kanellopoulos N. K.: 1999, Simulation of self-diffusion of point-like and finite-size tracers in stochastically reconstructed Vycor porous glasses. *J. Chem. Phys.* **111**, 2735.

Kim, I. C. and Torquato, S.: 1992, Diffusion of finite-sized Brownian particles in porous media. *J. Chem. Phys.* **96**, 1498.

Kirkpatrick, S., Gelatt C. D. and Vecchi, M. P.: 1983, Optimization by simulated annealing. *Science* **200**, 671.

Knight, R., Chapman, A. and Knoll, M.: 1990, Numerical modeling of microscopic fluid distribution in porous media. *J. Appl. Phys.* **68**, 994.

Leverett, M. C.: 1939, Flow of oil–water mixtures through unconsolidated sands. *Trans. AIME* **132**, 149.

Levitz, P., Ehret, G., Sinha, S. K. and Drake, J. M.: 1991, Porous Vucor glass – The microstructure as probed by electron-microscopy, direct energy-transfer, small-angle scattering, and molecular adsorption. *J. Chem. Phys.* **95**, 6151.

Levitz, P., and Tchoubar, D.: 1992, Disordered porous solids – from chord distributions to small angle scattering. *J. Phys. I* **2**, 771.

Lin, M. Y., Abeles, B., Huang, J. S., Stasiewski, H. E. and Zhang, Q.: 1992, Viscous flow and diffusion of liquids in microporous glasses. *Phys. Rev. B* **46**, 10701.

Makri, P. K., Romanos, G., Steriotis Th., Kanellopoulos, N. K. and Mitropoulos A. Ch.: 1998, Diffusion in a fractal system. *J. Colloid Interface Sci.* **206**, 605.

Manwart, C., Torquato, S and Hilfer, R.: 2000, Stochastic reconstruction of sandstones. *Phys. Rev. E* **62**, 893.

Martys N. S.: 1999, Diffusion in partially-saturated porous materials. *Materi. Struct.* **32**, 555.

Martys, N. and Garboczi, E. J.: 1992, Length scales relating the fluid permeability and electrical conductivity in random two-dimensional model porous media. *Phys. Rev. B* **46**, 6080.

Metropolis, N., Rosenbluth, A. W., Rosenbluth, M. N., Teller, A. H. and Teller, F.: 1953, Equations of state calculations by fast computing machines. *J. Chem. Phys.* **21**, 1087.

Mohanty, S.: 1997, Effect of multiphase fluid saturation on the thermal conductivity of geologic media. *J. Phys. D: Appl. Phys.* **30**, L80.

Papadopoulos, G.: 1993, Study of adsorption, diffusion and gas relative permeability in mesoporous alumina membranes, in relation to their porous and macroscopic structure, Ph.D. thesis (in Greek), University of Athens.

Pavlovitch, A., Jullien, R. and Meakin, P.: 1991, Geometrical properties of a random packing of hard spheres. *Physica A* **176**, 206.

Press W. H., Teukolsky S. A., Vetterling W. T., Flannery B. P.: 1992, Numerical Recipes in Fortran. Cambridge University Press.

Quiblier J. A.: 1986, New 3-dimensional modeling technique for studying porous media. *J. Colloid Interface Sci.* **98**, 84.

Reiss, H.: 1992, Statistical geometry in the study of fluids and porous media. *J. Phys. Chem.* **96**, 4736.

Roberts, J. N. and Schwartz, L. M.: 1985, Grain consolidation and electrical conductivity in porous media. *Phys. Rev. B* **31** 5990.

Rosanne, M., Mammar, N., Koudina, N., Prunet-Foch, B, Thovert, J.-F., Tevissen, E. and Adler P. M.: 2003, Transport properties of compact clays – II. Diffusion. *J. Colloid Interface Sci.* **260**, 195.

Schwartz, L. M., Garboczi, E. J. and Bentz, D. P.: 1995, Interfacial transport in porous media – Application to DC electrical conductivity of mortars. *J. Appl. Phys.* **78**, 5898.

Silverstein, D. L. and Fort, T.: 2000, Prediction of air–water interfacial area in wet unsaturated porous media. *Langmuir* **16**, 829.

Silverstein, D. L. and Fort, T.: 2000a, Incorporating low hydraulic conductivity in a numerical model for predicting air–water interfacial area in wet unsaturated particulate porous media. *Langmuir* **16**, 835.

Silverstein, D. L. and Fort, T.: 2000b, Prediction of water configuration in wet unsaturated porous media. *Langmuir* **16**, 839.

Stefanopoulos, K. L., Beltsios, K., Makri, P. K., Steriotis, T. A., Mitropoulos, A. Ch. and Kanellopoulos, N. K.: 2000, Characterization of the flow properties in Vycor by combining dynamic and scattering techniques. *Physica B* **276**, 477.

Tomadakis, M. M. and Sotirchos, S. V.: 1993, Ordinary and transition regime diffusion in random fiber structures. *AIChE J.* **39**, 397.

Visscher, W. M. and Bolsterli, M: 1972, Random packing of equal and unequal spheres in 2 and 3 dimensions. *Nature* **239**, 504.

Vold M. J.: 1960, The sediment volume in dilute dispersions of spherical particles. *J. Phys. Chem.* **64**, 1616.

Yao, J., Frykman, P., Kalaydjian, F., Thovert, J. F. and Adler, P M.: 1993, High order moments of the phase function for real and reconstructed model porous media. A comparison. *J. Colloid Interface Sci.* **156**, 478.

Yeong, C. L. Y. and Torquato, S.: 1998a, Reconstructing random media. *Phys. Rev. E* **57**, 495.

Yeong, C. L. Y. and Torquato S.: 1998b, Reconstructing random media. II. Three-dimensional media from two-dimensional cuts. *Phys. Rev. E* **58**, 224.

Transp Porous Med (2005) 58:63–86
DOI 10.1007/s11242-004-5470-8

Pore-Network Modeling of Isothermal Drying in Porous Media

A. G. YIOTIS[1,2], A. K. STUBOS[1,*], A. G. BOUDOUVIS[2],
I. N. TSIMPANOGIANNIS[3] and Y. C. YORTSOS[4]
[1]*National Center for Scientific Research "Demokritos"-IPTA, Ag. Paraskevi Attikis, 15310 Athens, Greece*
[2]*School of Chemical Engineering, National Technical University of Athens, 15780 Athens, Greece*
[3]*Los Alamos National Laboratory, Earth & Environmental Sciences Division (EES-6), Los Alamos, NM 87545, USA*
[4]*Department of Chemical Engineering, University of Southern California, Los Angeles, CA 90089-1211, USA*

(Received: 24 July 2003; in final form 22 April 2004)

Abstract. In this paper we present numerical results obtained with a pore-network model for the drying of porous media that accounts for various processes at the pore scale. These include mass transfer by advection and diffusion in the gas phase, viscous flow in the liquid and gas phases and capillary effects at the liquid–gas interface. We extend our work by studying the effect of capillarity-induced flow in macroscopic liquid films that form at the pore walls as the liquid–gas interface recedes. A mathematical model that accounts for the effect of films on the drying rates and phase distribution patterns is presented. It is shown that film flow is a major transport mechanism in the drying of porous materials, its effect being dominant when capillarity controls the process, which is the case in typical applications.

Key words: drying, films, corner flow, pore network

1. Introduction

Drying of porous media is a subject of significant scientific and applied interest. It is involved in the industrial drying of products such as food, paper, textile, wood, ceramics, granular and building materials, etc. Drying is also involved in distillation and vaporization processes associated with soil remediation (Ho and Udell, 1995), as well as in the recovery of volatile oil components from reservoirs by gas injection (Le Gallo et al., 1997).

In general, a single- or multi-component liquid phase gradually evaporates during drying and is removed from the porous structure via combined heat and mass transfer. Traditional descriptions of the process rely

---

*Author for correspondence: e-mail: stubos@ipta.demokritos.gr

on phenomenological approaches, in which the porous medium is a continuum, the dependent variables, like moisture content, are volume-averaged quantities and the relation of fluxes to gradients is through empirical coefficients. Such approaches essentially ignore the effect of the pore microstructure which is of key importance for a quantitative understanding of the process. Drying is a two phase flow process that involves many pore-scale mechanisms, for example the motion of individual gas–liquid menisci residing in the pore space, diffusion in the gas phase (for a single-component liquid) and the liquid phase (for a multi-component liquid), viscous flow in both phases, capillarity and liquid flow through connected films. All these mechanisms need to be accounted for at the pore scale.

Pore-network approaches are used extensively in recent years to model various processes in porous materials such as drying, immiscible two- and three-phase flow, solution gas-drive and many other (Li and Yortsos, 1995a; Valavanides and Payatakes, 2001; van Dijke et al., 2001). Pore network models describe processes at the pore- and pore-network scale and they offer better understanding of the physics involved in these processes than macroscopic continuum models that were used in the past.

Several studies used a pore-network approach to model drying in porous media in recent years. Key to these approaches is the consideration of mass transfer, elements of which were described by Li and Yortsos (1995b) and Jia et al. (1999), among others. Various pore-network models with specific applications to drying were proposed originally by Nowicki et al., and more recently in a series of papers by Prat and co-workers. Nowicki et al. (1992) presented a numerical simulation of the process at the pore-network level without expanding further on the particular patterns and regimes obtained or on the associated effects on drying rates. Prat's studies represent the first attempt to characterize theoretically drying patterns and their rate of change in porous structures. Prat (1995) studied drying patterns assuming capillary control, neglecting viscous effects and considering mass transfer only by quasi-static diffusion. Laurindo and Prat (1996, 1998) also provided a macroscopic assessment of the importance of liquid films that form at the pore walls as the liquid–gas interface recedes. Based on percolation patterns (Wilkinson and Willemsen, 1983) and isothermal conditions, they computed drying rates by solving a quasi-static diffusion equation in the gas phase. Prat and Bouleux (1999) focused on diffusional mass transfer and the effect of gravity on the front structure, but also commented on viscous effects.

In earlier experiments using horizontal glass-bead packs (Shaw, 1987), viscous forces were found to be important for explaining the formation of an evaporating front (separating continuous liquid from gas) of a finite size. More generally, we expect that advection and viscous effects will have an impact on patterns and drying rates. Existing pore-network models address mostly slow drying, controlled by capillarity and/or gravity and by diffusion, ignoring advection and/or viscous effects. They also neglect the role of liquid films in the process.

In the first part of this paper we present numerical results from a pore network simulator for the drying of porous media that accounts for all major mechanisms at the pore scale but ignores the effect of liquid films. We study mechanisms that have not been accounted for before such as viscous flow in both the gas and the liquid phases and the effect of viscous flow on the movement of the liquid–gas interface. A detailed description of this first part, including mathematical formulation, the algorithm and more extensive results can be found in a recent publication of the present authors (Yiotis et al., 2001). In the second part of the paper, the presence of liquid films is considered. We model capillarity-driven liquid flow in a 2D pore network and propose a mathematical model that accounts for viscous flow in the liquid films as well as all the other mechanisms presented earlier. This part is a detailed description of the study by Yiotis et al. (2003) (published elsewhere as a brief report) and an extension of that work.

2. Pore-Network Modeling of Drying without the Presence of Liquid Films

We consider the isothermal drying of a fractured porous medium initially saturated with a volatile liquid. The liquid is trapped in the pore space due to capillary forces and may vaporize as a result of an injected purge gas flowing primarily in the fractures. This process is applied to enhance oil recovery from reservoirs (Stubos and Poulou, 1999).

The actual overall problem is quite complex, requiring the consideration of the network of fractures and the medium continuum, gas flow and mass transfer in the fracture network and the multi-dimensional mass transfer from the medium continuum to the fracture network. For simplicity, we consider a 2D square pore network with all but one boundaries impermeable to flow and mass transfer (Figure 1).

At any time during the process, evaporation of the liquid at the liquid–gas interface leads to the receding of the liquid front (denoted as evaporating interface (I) or percolation front (P) in Figure 1), leaving behind disconnected clusters of liquid and liquid films, the size and location of which change continuously with time. In general, three different spatial regions can be identified:

(i) a far-field (from the fracture) region consisting of the initial liquid (continuous liquid cluster, CC);

(ii) a region where the liquid phase is disconnected and consists of individual liquid clusters of variable sizes (disconnected clusters, DC); and

(iii) a near-field (to the fracture) region consisting primarily of the continuous gas phase, with the liquid phase in the form of pendular rings,

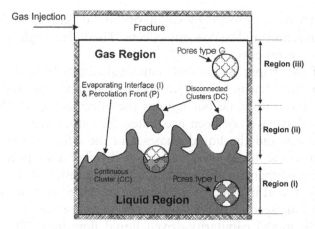

Figure 1. Schematic representation of a drying process in a 2D matrix driven by the injection of a purge gas through a fracture along the upper size of the matrix.

corner films or thin films on the solid surface, the thickness of which is progressively reduced towards a "totally dry" regime.

In Figure 1, the fracture running along the upper side of the 2D network is represented as a 1D chain of pore bodies and throats. The two ends of the fracture are open to flow and mass transfer. The network consists of spherical pore bodies connected through cylindrical pore throats. Pore bodies serve as containers for either of the two phases and it is assumed that they have no capillary or flow resistance. Therefore, when a liquid–gas interface lies within a pore body, the pressures of both phases are taken to be equal and the interface recedes without any capillary forces restraining its movement. The throats serve as conductors of the flow and mass transfer and they act as capillary barriers.

Initially the network is saturated by a single-component liquid (hexane). The fracture, however, contains only air at the beginning of the drying process. A purge gas (air) is injected at a constant volumetric rate from one end of the fracture. The concentration of the liquid component vapors is assumed to be zero at the entrance and the exit of the fracture at all times. The gas injection results to a pressure gradient along the fracture, which eventually develops inside the pore network as well. As a result of the gas flow, the liquid evaporates initially at the interface pore throats along the fracture. Vapor flows by advection and diffusion to the exit of the fracture.

Mass transfer of the vapor in the gas phase obeys the convection-diffusion equation

$$\frac{\partial C}{\partial t} + \mathbf{u} \cdot \nabla C = D \nabla^2 C \tag{1}$$

where C is the vapor concentration, D is the diffusion coefficient and $\mathbf{u}$ is the gas-phase velocity vector.

Both in the liquid and gas phases the fluxes Q between adjacent pores are computed by Poiseuille-law type flow resistances, where the viscosity μ is taken constant

$$Q_{ij} = \left(\frac{P_i - P_j}{\ell} \right) \frac{\pi r_{ij}^4}{8\mu} \qquad (2)$$

$$\sum_j Q_{ij} = 0 \qquad (3)$$

where P is the pressure at the center of each pore, r_{ij} is the throat radius between pores i and j and ℓ is the distance between pore centers.

Liquid menisci that reside at interface throats recede due to evaporation of the liquid and the gas invades the adjacent liquid pores. We assume that the menisci recede instantly because throats have no volume. At every liquid cluster, at least one liquid pore empties at any time during drying while all other liquid menisci may remain stationary due to capillary forces. A pore is invaded when the pressure difference across its throats exceeds the capillary pressure threshold $2\gamma/r$, where γ is the surface tension. If the pressure difference is not sufficiently large and all menisci of a cluster are pinned then the pore where the pressure difference is closer to its throats' capillary pressure threshold is invaded.

The problem is mainly characterized by two dimensionless parameters, a diffusion-based capillary number, Ca_D, and a Peclet number, Pe, in addition to the various geometrical parameters of the pore network

$$Ca_D = \frac{D\mu_1 C_e}{\gamma \ell \rho_1} \qquad (4)$$

$$Pe = \frac{V_f \ell}{d} \qquad (5)$$

where C_e is the equilibrium concentration of the vapor, V_f is the mean velocity of the purge gas in the fracture and the liquid phase is denoted by subscript 1. The capillary number expresses the ratio of viscous to capillary forces, based on a diffusion-driven velocity, while the Peclet number expresses the ratio of inertial to diffusion forces. We note that liquid films are neglected in this formulation. The details of the algorithm followed for the numerical solution of the problem can be found in Yiotis et al. (2001).

We present here two runs on a 50×50 pore network that are characteristic of the two limiting regimes that develop in this process. In the first run the gas flow rate through the fracture is very low and the process is characterized by a very low value for the Peclet number ($Pe = 0.66$ – run 15) (Figures 2 and 4). In this case capillary forces are dominant and mass transfer occurs primarily by diffusion. In the second run the purge gas is injected at a very high flow rate leading to a very high value of the Peclet number ($Pe = 331$ – run 4) (Figures 3 and 5). In this case viscous forces dominate at the liquid–gas interface while mass transfer occurs primarily by advection.

In the low Peclet number case, viscous forces are not sufficiently strong and the capillary pressure variation is negligible at the perimeter of liquid clusters. Assuming the absence of liquid films that could provide hydraulic conductivity between macroscopically disconnected liquid clusters (DCs) and the continuous liquid cluster (CC), every cluster takes the pattern of

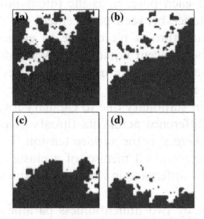

Figure 2. Phase distribution patterns for a low Peclet number ($Pe = 0.66$ – run 15). The process follows Invasion Percolation rules at all times. Air is white and liquid hexane is black.

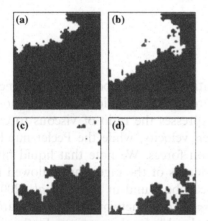

Figure 3. Phase distribution patterns for a high Peclet number ($Pe = 331$ – run 4). At early stages the process follows Invasion Percolation in a Stabilizing Gradient rules. Later on patterns become Invasion Percolation-like. Air is white and liquid hexane is black.

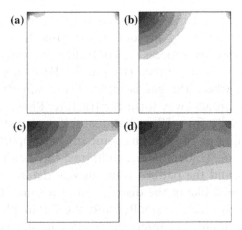

Figure 4. Concentration contours for a low Peclet number case ($Pe = 0.66$ – run 15). The snapshots correspond to the phase distribution patterns shown in Figure 2. Darker colors indicate lower vapor concentration.

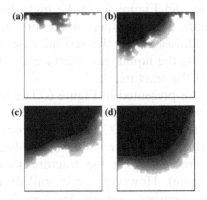

Figure 5. Concentration contours for the high Peclet number case ($Pe = 331$ – run 4). The snapshots correspond to the phase distribution patterns shown in Figure 3. Darker colors indicate lower vapor concentration.

Invasion Percolation (IP), in which the next throat to be invaded by the gas is that with the smallest capillary threshold (here, the one with the largest size) among all perimeter throats of that cluster (Figure 2). In the related study of Li and Yortsos (1995a, b) this process was termed *local percolation*.

Clusters closer to the open boundary are subject to a faster evaporation, compared to those further away, and are emptied faster (Figure 4). The end result is the development of gradients in the size of the isolated liquid clusters, with clusters closer to the fracture having smaller size. These gradients reflect mass transfer, rather than viscous effects. Clearly, however, the overall pattern would be a function of the drying rates, namely of the value of the *Pe*. We must note that under this regime of local percolation, different clusters may have different-size throats being invaded at the same time.

In the high Peclet number case the process is controlled by viscous forces and capillarity is negligible at early times (Figure 3). Pore throats closer to the inlet of the fracture, where the gas is injected, are subject to stronger viscous effects than pore throats away form the fracture. Phase distribution patterns deviate substantially from IP and almost follow a piston-like displacement (PD). The receding of the CC has some of the properties of IPSG in a fracture-matrix system. Under these conditions, the capillary resistance of a throat is negligible, and the pattern is exclusively determined by mass transfer considerations, much like in the dissolution of a solid. The rate of generation of DCs and their size are smaller and the liquid phase consists mostly of a CC. As the liquid–gas interface recedes in the pore network away from the fracture, viscous forces become weaker and the process gradually becomes of IP type, namely controlled by capillarity.

Concentration contours for these two limiting cases are shown in Figure 4 for the low Peclet number case and Figure 5 for the high Peclet number case. The concentration contours are smoother in the first case because mass transfer is controlled by diffusion. In the second case, the concentration contours are very steep close the liquid–gas interface and the vapor concentration is very low close to the fracture.

Typical drying curves for the process are presented in Figure 6. The high Peclet number case (*Pe* = 331) shows a clear constant rate period (CRP) for the drying rate that lasts approximately as long the CC remains in contact with the fracture. This CRP is much shorter for lower values of the Peclet number (i.e. *Pe* = 132) because the CC recedes form the fracture sooner (with respect to the liquid recovery volume). However, the overall drying rate is much shorter in the high Peclet number case as expected. More results and discussion including cases of intermediate Peclet numbers are shown in Yiotis *et al.* (2001).

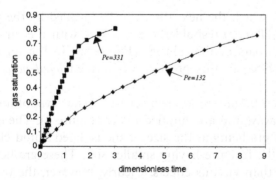

Figure 6. Drying curves for various Peclet numbers.

3. The Effect of Liquid Films

In this section we study the role of wetting films in the context of drying. The objective is to determine if film flows can be a significant mechanism of drying as purported to be in the experiments cited below. A 2D pore-network model, very similar to the one described above, is used for the representation of the porous medium. In this case however we consider that the velocity of the purge gas in the fracture is very low and that viscous forces due to flow in the gas phase on the liquid–gas interface are negligible. Our focus is on the effect of viscous flow through the liquid films that develop at the corners of the pore network.

Flow through macroscopic films has been analyzed in the context of imbibition by many authors (Lenormand and Zarcone, 1984; Lenormand, 1992; Dullien et al., 1989; Constantinides and Payatakes, 2000). Lenormand (1992) described in detail the expected mechanisms due to film flow in imbibition. Dullien et al. (1989) reported flow along surface microchannels that provide hydraulic connectivity between macroscopically isolated liquid regions during immiscible displacement in packed glass beads with rough surfaces. This was found to be negligible in the case of smooth glass beads. Dillard and Blunt (2000) examined mass transfer from liquid films in dissolution processes, while Blunt et al. (2002) presented a detailed review of flow through films in the context of three-phase flow.

In the context of drying, past experimental work has emphasized the existence and speculated on the role of film flow (Shaw, 1987; Laurindo and Prat, 1998; Tsimpanogiannis et al., 1999). In a series of experiments Shaw (1987) found that, under comparable conditions, the drying front in a cell containing packed beads moved one order of magnitude faster than when the cell was empty. Shaw attributed this "unorthodox" result to liquid counterflow through films which form along particle contacts, and argued that it is the dominant mechanism for the drying of porous materials. Laurindo and Prat (1998) performed drying experiments in two-dimensional etched-glass micromodels and compared their results with predictions from a pore-network simulator, which did not contain films. The experimental rates were found to be about six times higher than the numerical. These authors also attributed the flow enhancement to wetting liquid films and presented a simplified model for the associated transport mechanism. Liquid films were described in the form of a bundle of microcapillaries on the pore surface. However, no quantitative models for film flow were developed in these studies.

3.1. PROBLEM FORMULATION

We consider the isothermal evaporation of a single-component liquid in a porous medium one side of which is open to the environment. The latter is kept dry through the flow of a purge gas, while all other three sides are impermeable to fluid flow and mass transfer. The porous medium is again represented by a 2D square lattice of pores connected through throats with square cross-section. The radii of the throats, hence the corresponding capillary thresholds, are distributed randomly. In the presence of films, at any stage of drying, the pore space can be characterized by three kinds of pores (close-ups in Figure 7): pores L, fully occupied by liquid, pores G, fully occupied by gas, and pores F, occupied by gas but also containing liquid films. The existence of pores of type F is the distinguishing feature of this work, compared to previous (Prat, 1995; Laurindo and Prat, 1998; Prat and Bouleux, 1999; Tsimpanogiannis et al., 1999; Yiotis et al., 2001, Plourde and Prat, 2003). Our focus is on thick films, e.g. which form in the corners of polygonal pores, and where flow is driven by capillary pressure gradients. Here, we will account for viscous effects both in the films (F pores), as well as in the continuous liquid phase (region L). Mass transfer in the gas phase is assumed only by diffusion, which is usually valid in drying problems (Prat, 1995; Laurindo and Prat, 1996, 1998).

Due to the applied concentration gradients, the liquid evaporates along the liquid–gas interfaces and the liquid vapors are transferred by diffusion in the gas phase towards the dry end. During this process, the macroscopic gas–liquid interfaces (denoted by P in Figure 7) recede, both in the continuous and the discontinuous clusters.

3.2. LIQUID FLOW THROUGH MACROSCOPIC FILMS IN A SINGLE CAPILLARY

We first study the case of long capillary with square cross-section where the liquid meniscus has just receded leaving behind liquid films at the four corners of the capillary (Figure 8). The thickness of each liquid film can

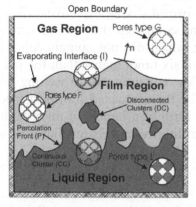

Figure 7. Schematic of liquid and gas phase patterns, indicating the various types of pores in drying used in this study.

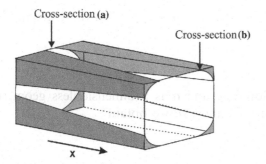

Figure 8. Liquid films along the corners of a capillary with square cross-section.

be parameterized by its radius of curvature r, which is a function of time and distance. We assume that all four films have the same thickness at any cross-section along the capillary. Assuming local capillary equilibrium at the film interface, we have

$$P_c = P_g - P_l = \frac{\gamma}{r} \tag{6}$$

By neglecting variations in the gas pressure, we can take without loss, $P_g = 0$. Then, the pressure in the film, P_l, is

$$P_l = -\frac{\gamma}{r} \tag{7}$$

According to Equation (7) the liquid pressure in the film is inversely proportional to its thickness. Any gradient in the film thickness along the capillary results in a pressure gradient along the liquid films. A capillarity-induced flow develops along the film from the cross-section where the film is thicker (Figure 9(a)) towards the cross-section where the film is thinner (Figure 9(b)).

Several authors (Lenormand and Zarcone, 1984; Ransohoff and Radke, 1988; Zhou *et al.*, 1997) have studied film flow along the corners of long smooth capillaries with polygonal cross-sections. Assuming uni-directional

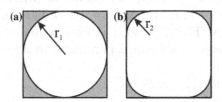

Figure 9. Evolution of the film radius that develops at the corners of a capillary with square cross-section. Cross-section (a) is closer to the bulk liquid phase and the films are thicker than in cross-section (b).

(direction x) viscous flow, a Poiseuille-type law applies in these geometries,

$$Q_x = -\frac{\alpha r^4}{\mu_1} \frac{\partial P_1}{\partial x} \tag{8}$$

where Q_x is the volumetric flow rate and α is a dimensionless geometric factor. By combining Equations (7) and (8), we obtain

$$Q_x = -\frac{\alpha \gamma r^2}{\mu_1} \frac{\partial r}{\partial x} \tag{9}$$

Parameter α was determined previously for various model geometries (Ransohoff and Radke, 1988, Dong and Chatzis, 1995). The latter authors studied film flow in one corner of a capillary with a polygonal cross-section and found the following expression

$$\alpha = \frac{C^*}{\beta} \quad \text{where } C^* = 4\left(\frac{\cos\theta\cos(\pi/4+\theta)}{\sin(\pi/4)} - (\pi/4 - \theta)\right) \tag{10}$$

the shape factor C^* being expressed in terms of the contact angle θ. The dimensionless resistance β was earlier calculated by Ransohoff and Radke (1988). In the case of a capillary with square cross-section and a perfect wetting liquid ($\theta = 0$) we have $C^* = 4 - \pi$, $\beta = 93.5$ and $\alpha = 0.0088$. We note that parameter α is of order 10^{-3}.

We consider the mass balance for the evaporating liquid in the capillary

$$(4 - \pi)\frac{\partial r^2}{\partial t} = -\frac{\partial Q_x}{\partial x} - Q_{ev} \tag{11}$$

where Q_{ev} is the evaporation rate. Assuming a simple diffusion model we take

$$Q_{ev} = \left(\frac{2\pi r D}{\rho_1}\right)\frac{(C_e - C)}{r_0} \tag{12}$$

where C is the average gas-phase mass concentration (mass per unit volume) of the evaporating liquid and r_0 is the radius of curvature where the films emanate. The particular details of film evaporation, particularly near the tip of the film are very complex. Equation (12) is only a first-order approximation, and the value of D acts as a lumped parameter to account for many of the neglected phenomena.

Combining Equations (9), (11) and (12) leads to

$$\frac{\partial r}{\partial t} = \frac{2\alpha\gamma}{(4 - \pi)\mu_1}\left[r\frac{\partial^2 r}{\partial x^2} + 2\left(\frac{\partial r}{\partial x}\right)^2\right] - \frac{\pi D}{(4 - \pi)\rho_1}\frac{(C_e - C)}{r_0} \tag{13}$$

For dimensionless notation, we introduce the diffusive time $\tau = \frac{Dt}{r_0^2}$, a rescaled radius of curvature, $\rho = \frac{r}{r_0}$, a rescaled axial distance, $\xi = \frac{x}{r_0}$, and a dimensionless concentration $\zeta = \frac{C}{C_e}$ based on which we write

$$\left(\frac{(4-\pi)\rho_l}{\pi C_e}\right) Ca_F \frac{\partial\rho}{\partial\tau} = \rho\frac{\partial^2\rho}{\partial\xi^2} + 2\left(\frac{\partial\rho}{\partial\xi}\right)^2 + Ca_F(\zeta-1) \tag{14}$$

Here we introduced the capillary number in the form $Ca_F = \frac{\pi D C_e 2\mu_l\beta}{\rho_l C^* r_0\gamma}$. As in other mass-transfer driven processes, this capillary number expresses the ratio of the viscous forces due to flow driven by mass transfer to capillary forces.

The mass balance in the gas phase reads

$$\frac{\partial C}{\partial t} = D\frac{\partial^2 C}{\partial x^2} + \frac{1}{2}\frac{\pi r D}{r_0^3}(C_e - C) \tag{15}$$

and in dimensionless notation,

$$\frac{\partial\zeta}{\partial\tau} = \frac{\partial^2\zeta}{\partial\xi^2} + \frac{1}{2}\pi\rho(1-\zeta) \tag{16}$$

Equations (14) and (16) are coupled and solved for the dimensionless film thickness ρ and the dimensionless concentration ζ as a function of time τ and capillary length ξ with the appropriate initial and boundary conditions. We find that for realistic values of the capillary number Ca_F, a steady-state profile for ρ and ζ is quickly established (Figure 10). This is consistent with other diffusion-controlled problems in porous media (Witten and Sanders, 1981; Peitgen and Saupe, 1988). In addition, the restriction of the evaporation to a narrow region is a consequence of the exponential decay of the concentration: Due to the confined pore geometry, the gas phase becomes rapidly saturated, limiting evaporation to a narrow region near the film tips, where the liquid flux is supplied by capillarity-driven film flow (Yiotis et al., 2003).

Given that $\zeta = 1$ along the film region and that a steady-state profile for ρ is quickly established, Equation (14) can be simplified as follows

$$\rho\frac{\partial^2\rho}{\partial\xi^2} + 2\left(\frac{\partial\rho}{\partial\xi}\right)^2 = 0 \Rightarrow \frac{\partial^2\rho^3}{\partial\xi^2} = 0 \tag{17}$$

In the dry region, between the film tip and the open end of the capillary, the film thickness is $\rho = 0$ and liquid vapors are transferred by diffusion towards the open end of the capillary. Equation (16) becomes

$$\frac{\partial^2\zeta}{\partial\xi^2} = 0 \tag{18}$$

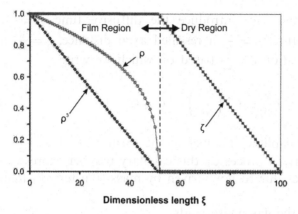

Figure 10. Steady-state profiles for the dimensionless film thickness ρ and vapor concentration ζ along a single capillary. The film emanates at $\xi = 0$.

We note that Equations (17) and (18) apply along the film and in the totally dry part of the capillary, respectively. The mass balance at the film tip reads

$$\frac{\pi}{6Ca_F}\frac{\partial\rho^3}{\partial\xi} = \frac{\partial\zeta}{\partial\xi} \quad \text{and} \quad \rho = 0, \quad \zeta = 1 \tag{19}$$

Given that the location of film tip is unknown, we consider a simple transformation that leads to a rather straightforward solution.

$$\Phi \equiv \frac{\rho^3 + \zeta Ca_F}{1 + Ca_F} \tag{20}$$

Function Φ satisfies Equations (17) and (18) in their domains and is continuous at the film tip. Equations (17) and (18) can be replaced by

$$\frac{\partial^2\Phi}{\partial\xi^2} = 0 \tag{21}$$

The location of the film tip is calculated from the solution of Equation (21) by applying the appropriate boundary conditions at the beginning of the film and the open end of the capillary. Note that in our model the distance of the film tip from the open end of the capillary actually determines the effect of films in the drying process. The closer is the film tip to the open end of the capillary the higher is the drying rate.

3.3. DRYING THROUGH MACROSCOPIC FILMS IN A PORE NETWORK

The single-capillary physics described above are also expected to apply to the general problem, where films exist in a network of pores connected through capillaries similar to the one considered above. As described in Section 3.1, the gas region of such a network contains F pores adjacent to capillaries that contain films (film region) and G pores adjacent to dry capillaries (dry region) (Figure 7). Given that Equation (17) applies in the film region of a single capillary, we can assume that the film thickness in the film region of a 2D pore network is described by the following Laplace equation

$$\nabla^2 \rho^3 = 0 \tag{22}$$

Assuming a quasi-steady state in the concentration in dry region and evaporation occurring where the films terminate, namely at the interface I between pores F and G, the vapor concentration in the dry region is described by the following Laplace equation

$$\nabla^2 \zeta = 0 \tag{23}$$

The mass balance at the film tips where the liquid evaporates (interface I in Figure 7) reads

$$\frac{\pi}{6 C a_F} \frac{\partial \rho^3}{\partial n} = \frac{\partial \zeta}{\partial n} \quad \text{and} \quad \rho = 0, \ \zeta = 1 \tag{24}$$

The film region is saturated with liquid vapors, $\zeta = 1$, and the film thickness is zero in the dry region, $\rho = 1$.

   Assuming that we know the location of the percolation front P at any time, we can solve the full problem using the transformation proposed in Section 3.2. Equations (22) and (23) become

$$\nabla^2 \Phi = 0 \text{(in regions G and F)} \tag{25}$$

We assume that the film thickness ρ is approximately constant at the percolation front P

$$\Phi = 1 \text{ at the percolation front P} \tag{26}$$

and that drying is driven by imposing $\zeta = 0$ at the open side of the network

$$\Phi = 0 \text{ at the open end of the network} \tag{27}$$

Using the above transformation, the solution of the Laplace equation (25) can be used to determine the profiles of the film radius and the concentration, the rates of drying through each film, as well as the location of the interface I, where the films terminate and evaporation occurs. Interface I is located simply as the position where

$$\Phi = \frac{C a_F}{C a_F + 1} \tag{28}$$

The location of the percolation front P is determined by solving for the liquid fluxes in the liquid phase and films, simultaneously. In all pores occupied fully by liquid (pores of type L), the viscous flow is described by
Poiseuille-type expressions. For these pores, the liquid mass balance at any
pore i reads in dimensionless notation

$$\sum_j (p_i - p_j)\sigma_{ij}^4 = 0 \tag{29}$$

where j denotes a neighboring pore, σ is the normalized pore radius, and
we have normalized pressure drops with the characteristic value $P^* = \frac{\gamma}{r_0}$.
For pores at the perimeter P, however, the mass balance is different and
we need to further distinguish two cases.

If the pressure difference at a pore that belongs to the percolation front
is not sufficiently large for the gas phase in the neighboring pore to penetrate the connecting throat, the meniscus remains stationary. Because of the
films, however, there is always net liquid outflow from the liquid pore. The
mass balance at every such pore at the percolation front reads

$$\sum_j (p_P - p_j)\sigma_{ij}^4 = -\frac{32\alpha}{3\pi}(1 + Ca_F)\sum_F \left.\frac{\partial \Phi}{\partial \xi}\right|_P \tag{30}$$

where the first sum denotes the liquid arriving at the perimeter pore P and
the last sum denotes the contributions from the films in pores of type F
draining the perimeter pore.

If the capillary pressure at a perimeter throat is sufficiently high, namely
if the following condition is satisfied

$$-p_P > \frac{1}{\sigma_{ij}} \tag{31}$$

the adjacent pore is penetrated by the gas phase. As in Yiotis et al. (2001),
in such pores we assume that the capillary pressure is zero. Then, the corresponding mass balance reads in dimensionless notation

$$\sum_j p_j \sigma_{ij}^4 - \frac{32\alpha}{3\pi}(1 + Ca_F)\sum_F \left.\frac{\partial \Phi}{\partial \xi}\right|_P = \frac{8\mu_1 Q_P}{\pi r_0^2 \gamma} \tag{32}$$

where we have taken $p_P = 0$. Q_P is the flow rate at which the pore empties.

Equations (30) and (32) are solved for the pressure field in the liquid
phase and the flow rates Q_P that penetrated pores empty given the solution of the Laplace equation (25) in the gas phase that determines the liquid fluxes through the liquid films at the perimeter of the liquid clusters.

The gas saturation at every pore that has been penetrated by the gas phase is calculated explicitly in time

$$\Delta S_i = \frac{\Delta t}{V_P} \cdot Q_P \qquad (33)$$

where V_P is the pore volume, Δt is the time step and Q_P is assumed constant during the time step.

The algorithm used in our simulator is based on the work by Yiotis *et al.* (2001) and can be summurized as follows: At any given time, pores have one of the designations L, F or G (Figure 7). The L pores can be part of the original liquid cluster (CC) or of the disconnected finite-size clusters (DC). At every time step, the overall rate of evaporation from each of the liquid clusters is evaluated using the Laplace equation (25). Pressure fields in the liquid clusters are calculated with Equations (30) and (32) and Partly Empty pores at the percolation front are emptied according to Equation (30). The time step is selected such that it equals the minimum time required to empty completely any of the available Partly Empty pores. If at the current time no Partly Empty pores are available to any (or all) of the clusters (namely if all pores are of the CE type, Yiotis *et al.*, 2001), the throat with the smallest capillary threshold in the perimeter of any given cluster is invaded next, at which time the invaded pore becomes of the PE type. To determine this throat, the liquid pressure is lowered uniformly in space inside the cluster, until the capillary pressure exceeds for the first time the smallest capillary threshold. Invasion must occur, since due to evaporation there is a continuous loss of mass from the liquid clusters through film flow. Then, the interfaces are updated, the equation for Φ is solved again, the rates of flow through the film obtained and the process continues. All calculations are done explicitly in time. The fields for the pressure and Φ are computed using Successive Over-Relaxation.

3.4. NUMERICAL RESULTS AND DISCUSSION

We performed a series of numerical simulations in 100×100 pore networks for different values of Ca_F to study the effect of liquid films on the extent of the liquid films, the phase distribution patterns and drying rates. The pore network consists of spherical pores connected through throats with square cross-section. All pores have the same radius $500 \, \mu$m while the size of the throats follows a random distribution between 170 and $270 \, \mu$m. We assume that the film flow occurs practically only in the throats that also act as capillary barriers. Pores serve as containers of either of the two phases

3.4.1. *Phase Distribution Patterns*

Equations (30) and (32) show that the pressure drop at the percolation front P (at the perimeter of liquid clusters) scales as $(1 + Ca_F)$. For values of Ca_F less than order of 1 (which is the typical case in most physical problems) we expect that the effect of the capillary number on the phase distribution patterns is negligible. In that range, phase distribution patterns follow Invasion Percolation rules. The left panel in Figure 11 shows two snapshots of the percolation front for $Ca_F = 10^{-4}$ that correspond to IP patterns.

As the capillary number increases, the patterns eventually depart from IP, particularly at the early times of the process. However, Figure 11 shows that quite large capillary number values are needed for a noticeable effect on the pattern. For example, the middle panel of Figure 11 shows that even for $Ca_F = 1$ the pattern is almost identical to IP, except for a few small differences at the start of the process. It takes larger values, of the order of $Ca_F = 10$ (right panel of Figure 11), for the effect to be pronounced. Then, the pressure drop at the front becomes relatively significant, and the pattern exhibits the expected behavior of viscous "stabilization" (Tsimpanogiannis *et al.*, 1999), as Invasion Percolation in a Stabilizing Gradient (IPSG)).

As drying progresses, the recovery rate diminishes due to the receding of the percolation front. In the large Ca_F case, this results to a transition from IPSG patterns to capillary-dominated IP patterns. It follows that under typical conditions and for all practical purposes, the drying front can be accurately described as an IP front. This is in contrast to the behavior of external drainage, where viscous effects on the pattern become important for values of the capillary number as low as 10^{-3}.

3.4.2. *Extent of Liquid Films*

The film properties are determined from the solution of the Laplace equation (25) for Φ. Figure 12 shows a snapshot of the iso-potential contours of Φ around the percolation front of the liquid clusters. The patterns reflect the solution of the Laplace equation around a fractal object, hence following the fractal features at distances close to it, but becoming smooth at a small distance away. Because of the assumed boundary condition $\Phi = 1$ at the cluster perimeter, all clusters act as sinks, therefore there is no fluid flow from one cluster to another.

For low values of Ca_F, which is typically the case, the films extend all the way to the open boundary (where $\Phi = 0$). By contrast, when Ca_F is of order 1, the films are short and the film tips (evaporation interface I) reside closer to the liquid cluster interface P (where $\Phi = 1$).

Figure 13 shows the evolution of the film radius profiles for two values of Ca_F. The film radius has the approximate profile $\rho \propto \xi^{1/3}$ where ξ is the

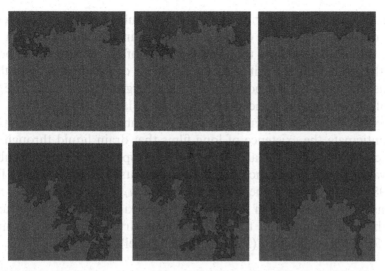

Figure 11. Two snapshots of the drying front for $Ca = 10^{-4}$ (left), $Ca = 1$ (middle) and $Ca = 10$ (right). Liquid-occupied pores are in red.

distance from the front, and which corresponds to the solution of Equation (25), as discussed above. The location of the film tips (interface I) is the contour with the value $\Phi = \frac{Ca_F}{Ca_F+1}$. Figure 13 shows that at low Ca_F (left panel), the films extend all the way to the open end, which is the place where practically all evaporation occurs. When Ca_F is of order 1 or larger, however, the films are shorter (right panel in Figure 13), and lead to the formation of a completely dry region G, the extent of which increases with

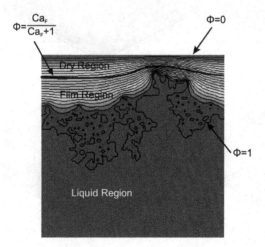

Figure 12. Iso-contours of the solution of the Laplace equation around the liquid clusters, for $Ca = 0.5$, with boundary conditions $\Phi = 1$ at the front and $\Phi = 0$ at the open end.

time. Even though the film region is short and a fully-dry region has developed, the drying front is still of the IP type (as in the left panel). In the O(1) case, the film tips mimic the protuberances of the drying front, being closer to the open boundary if associated with a corresponding protuberance. Films that end at such points will have a larger drying rate, since the gradient of Φ (and of the concentration) there is larger, (Figure 14).

In the typical problem, where the capillary number is generally less than O(1), we anticipate the existence of long films that drain liquid through the above "wicking" action, and all the way to the open end where it subsequently evaporates. For instance, in the experiments in Laurindo and Prat (1996, 1998) we have made the rough estimate $Ca_F \sim 10^{-4}$ suggesting that liquid films (film region F in Figure 7) likely existed in all gas-invaded pores in the experiments and that a completely dry region (gas region G in Figure 7) did not develop, (Figure 13, left panel).

3.4.3. *Drying Curves*
Based on the analysis in Section 3.3 the overall drying rate at the open side of the pore network is

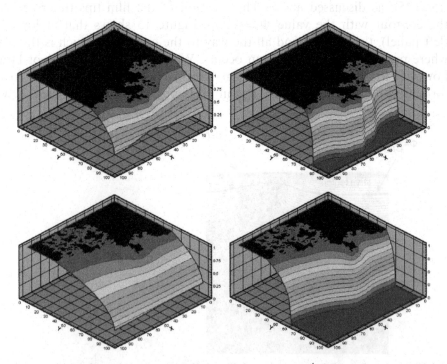

Figure 13. Profiles of the rescaled film radii for $Ca_F = 10^{-4}$ (left) and $Ca_F = 1$ (right) at two different stages of the process. Liquid clusters are in black, the fully dry region is in blue.

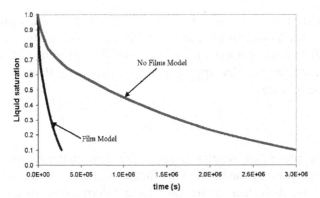

Figure 14. Drying curves calculated with a numerical model that does not account for the effect of liquid films and a model that accounts for liquid films for $Ca_F = 0.1$ under comparable conditions.

$$F = \frac{DC_e r_t^2}{\ell} \frac{1 + Ca_F}{Ca_F} F_D \tag{34}$$

where r_t is the average throat radius along the open boundary, $F_D = -\int_{a_0} \frac{\partial \Phi}{\partial n} da$ is the dimensionless drying rate and subscript 0 denotes the open boundary. The dimensionless rate F_D depends on the geometry of the porous medium.

Equation (34) shows that the drying rate scales as $\frac{Ca_F + 1}{Ca_F}$. The drying rate increases as Ca_F decreases and the film tips are closer to the open boundary. At smaller values of Ca_F, capillarity helps to transport liquid over larger distances and to keep the film extent longer. This is favored by larger interfacial tension, larger values of the film thickness r_0 at the percolation front (where films emanate) and smaller viscosity and effective diffusivity. It is readily shown that in the region of small Ca_F, the drying rate scales as

$$F \sim \frac{\gamma r_0}{\mu_1} \tag{35}$$

showing the dominant effect of capillarity in this region. Conversely, at values of Ca_F of order 1 and larger, the film extent is smaller, films do not contribute substantially, and the drying rates are smaller. There we have roughly,

$$F \sim \frac{D r_t^2 C_e}{\ell} \tag{36}$$

All previous pore-network models (Prat, 1995; Laurindo and Prat, 1996, 1998; Prat and Bouleux, 1999; Yiotis *et al.*, 2001) correspond effectively to such condition.

Figure 14 shows a drying curve calculated using a pore-network model that does not account for liquid films and a drying curve calculated with our model for $Ca_F = 0.1$ under comparable conditions. The presence of liquid films increases the drying rate by approximately a factor of 10 even for a relatively large value of Ca_F.

4. Conclusions

In this work, we first presented results from a 2D pore-network model for isothermal drying in porous media that includes mechanisms like mass transfer by advection and diffusion in the gas phase, viscous flow in liquid and gas phases and capillary effects at the gas–liquid menisci in the pore throats. In a further step, we proceeded to study the effect of capillarity-driven flow in macroscopic liquid films during the drying process. A mathematical model that accounts for viscous flow both through the liquid films and the bulk liquid phase was developed. Using a novel transformation, it was found that film flow is a major transport mechanism, its effect being dominant when capillarity controls the process, which is the case in typical applications.

We have shown that capillarity-induced flow through the films that form in cavities at the pore walls is favored by larger interfacial tension, larger values of the film thickness at the percolation front (where the films emanate) and smaller liquid viscosity. In typical drying problem the extent of the liquid films is approximately proportional to the surface tension. The liquid is transferred from the liquid clusters through the films towards the film tips where it evaporates. The longer is the film region the closer to the open boundary the liquid is transferred and the higher is the drying rate. Our results are in qualitative agreement with previous experimental work which shows accelerated drying when films contribute to flow.

The approach we followed was subject to several simplifying assumptions that may preclude the precise quantitative comparison with experimental results. However, we believe that we have provided a good first approximation to the complicated problem of film flow in drying processes.

References

Blunt, M. J., Jackson, M. D., Piri, M. and Valvante, P. H.: 2002, Detailed physics, predictive capabilities and macroscopic consequences for pore-network models of multiphase flow, *Adv. Water Resour.* **25**, 1069–1089.

Constantinides, G. N. and Payatakes, A. C.: 2000, Effects of precursor wetting films in immiscible displacement through porous media, *Trans. Porous. Med.* **38**, 291–317.

Dillard, L. A. and Blunt, M. J.: 2000, Development of a pore network simulation model to study nonaqueous phase liquid dissolution, *Water Resour. Res.* **36**, 439–454.

Dong, M. and Chatzis, I.: 1995, The imbibition and flow of a wetting liquid along the corners of a square capillary tube, *J. Colloid Interf. Sci.* **172**, 278–288.

Dullien, A. L., Zarcone, C., MacDonald, I. F., Collins, A. and Bochard, D. E.: 1989, The effects of surface roughness on the capillary pressure curves and the heights of capillary rise in glass bead packs, *J. Colloid Interf. Sci.* **127**, 362–772.

Ho, C. K. and Udell, K. S.: 1995, Mass transfer limited drying of porous media containing an immobile binary liquid mixture, *Int. J. Heat Mass Transfer* **38**, 339–350.

Jia, C., Shing, K. and Yortsos, Y. C.: 1999, Visualization and simulation of non-aqueous phase liquids solubilization in pore networks, *J. Contam. Hydrol.* **35**, 363–387.

Laurindo, J. B. and Prat, M.: 1996, Numerical and experimental network study of evaporation in capillary porous media. Phase distributions, *Chem. Eng. Sci.* **51**, 5171–5185.

Laurindo, J. B. and Prat, M.: 1998, Numerical and experimental network study of evaporation in capillary porous media. Drying rates, *Chem. Eng. Sci.* **53**, 2257–2269.

Le Gallo, Y., Le Romancer, J. F., Bourbiaux, B. and Fernades, G.: 1997, Mass transfer in fractured oil reservoirs during gas injection, SPE J 38924.

Lenormand, R.: 1992, Liquids in porous media, *J. Phys.* **2**, SA79–SA88.

Lenormand, R. and Zarcone, C.: 1984, Role of roughness and edges during imbibition in square capillaries, SPE J 13264.

Lenormand, R. and Zarcone, C.: 1985, Invasion percolation in an etched network: measurement of a fractal dimension, *Phys. Rev. Lett.* **54**, 2226–2229.

Li, X. and Yortsos, Y. C.: 1995a, Visualization and simulation of bubble growth in pore networks, *AIChE J.* **41**, 214–222.

Li, X. and Yortsos, Y. C.: 1995b, Theory of multiple bubble growth in porous media by solute diffusion, *Chem. Eng. Sci.* **50**, 1247–1271.

Luikov, A. V.: 1966, Heat and Mass Transfer in Capillary-Porous Bodies. Pergamon Press.

Nowicki, S. C., Davis, H. T. and Scriven, L. E.: 1992, Microscopic determination of transport parameters in drying porous media, *Drying Tech.* **10**, 925–946.

Peitgen, H.-M. and Saupe, D. (eds.): 1988, *The Science of Fractal Images*, Springer-Verlag.

Prat, M.: 1995, Isothermal drying of non-hygroscopic capillary-porous materials as an invasion percolation process, *Int. J. Multiphase Flow* **21**, 875–892.

Prat, M. and Bouleux, F.: 1999, Drying of capillary porous media with a stabilized front in two dimensions, *Phys. Rev. E* **60**, 5647–5656.

Ransohoff, T. C. and Radke, C. J.: 1988, Laminar flow of a wetting liquid along the corners of a predominantly gas-occupied noncircular pore, *J. Colloid Interf. Sci.* **121**, 392–401.

Shaw, T. M.: 1987, Drying as an immiscible displacement process with fluid counterflow, *Phys. Rev. Lett.* **59**, 1671–1674.

Stubos, A. K. and Poulou, S.: 1999, Oil recovery potential from fractured reservoirs by mass transfer processes, SPE J 56415.

Tsimpanogiannis, I. N., Yortsos, Y. C., Poulou, S., Kanellopoulos, N. and Stubos, A. K.: 1999, Scaling theory of drying porous media, *Phys. Rev. E* **59**, 4353–4365.

Valavanides, M. S. and Payatakes, A. C.: 2001, True-to-mechanism model of steady-state two-phase flow in porous media, using decomposition into prototype flows, *Adv. Water Resour.* **24**, 385–407.

van Dijke, M. I. J., Sorbie, K. S. and McDougall, S. R.: 2001, Saturation-dependencies of three-phase relative permeabilities in mixed-wet and fractionally wet systems, *Adv. Water Resour.* **24**, 365–384.

Wilkinson, D. and Willemsen, J. F.: 1983, Invasion percolation: a new form of percolation theory, *J. Phys. A* **16**, 3365–3370.

Witten, T. A. and Sander, L. M.: 1981, Diffusion-limited aggregation. A kinetic critical phenomenon, *Phys. Rev. Lett.* **47**, 1400–1403.

Yiotis, A. G., Stubos, A. K., Boudouvis, A. G., Tsimpanogiannis, I. N. and Yortsos, Y. C.: 2003, Effect of liquid films on the isothermal drying of porous media, *Phys. Rev. E*, **68**, 037303.

Yiotis, A. G., Stubos, A. K., Boudouvis, A. G. and Yortsos, Y. C.: 2001, A 2-D pore-network model of the drying of single-component liquids in porous media, *Adv. Water Resour.* **24**, 439–460.

Yortsos, Y. C. and Stubos, A. K.: 2001, Phase change in porous media, *Curr. Opin. Colloid Interf. Sci.* **6**, 208–216.

Zhou, D., Blunt, M. and Orr Jr., F. M.: 1997, Hydrocarbon drainage along corners of non-circular capillaries, *J. Colloid Interf. Sci.* **187**, 11–21.

Transp Porous Med (2005) 58:87–119
DOI 10.1007/s11242-004-5471-7

Phenomenological Meniscus Model for Two-Phase Flows in Porous Media

M. PANFILOV* and I. PANFILOVA
Laboratoire Environnement, Géomécanique & Ouvrages, ENS de Géologie, INP de Lorraine, Vandøeuvre-lès-Nancy, France

(Received: 7 November 2003; in final form: 15 April 2004)

Abstract. A new macroscale model of a two-phase flow in porous media is suggested. It takes into consideration a typical configuration of phase distribution within pores in the form of a repetitive field of mobile menisci. These phase interfaces give rise to the appearance of a new term in the momentum balance equation, which describes a vectorial field of capillary forces. To derive the model, a phenomenological approach is developed, based on introducing a special continuum called the *Meniscus-continuum*. Its properties, such as a unique flow velocity, an averaged viscosity, a compensation mechanism and a duplication mechanism, are derived from a microscale analysis. The closure relations to the phenomenological model are obtained from a theoretical model of stochastic meniscus stream and from numerical simulations based on network models of porous media. The obtained transport equation remains hyperbolic even if the capillary forces are dominated, in contrast to the classic model which is parabolic. For the case of one space dimension, the analytical solutions are obtained, which manifest non-classical effects as double displacement fronts or counter-current fronts.

Key words: capillary, two-phase flow, meniscus, hyperbolic equation, displacement, flow model, shocks, non-equilibrium.

1. Introduction

1.1. TWO-PHASE STRUCTURES IN POROUS MEDIA

At a fixed phase saturation, a two-phase flow through porous media may be characterized by various geometrical structures of phase distribution within the pore space. The basic examples of phase structures are: (a) a film: both the phases are present in each pore and are separated by a large-scale interface, whose normal vector is perpendicular, in average, to the mean flow direction in each channel; (b) a meniscus structure: both the phases are present in a number of pores and are separated by an interface of a pore scale; (c) a channel structure: each phase occupies its own

---

*Author for correspondence: e-mail: michel.panfilov@ensg.inpl-nancy.fr

pores forming continuous channels crossing the overall medium; the interface between phases is absent or minimal. The basic difference between these structures is the form and the dimension of the interface.

Therefore, a phase distribution in a porous medium is characterized by two (at least) macroscale parameters: the phase saturation s and a configuration parameter η, responsible for the interior structure of phase distribution. In general, the traditional parameters used to describe two-phase flow as the relative permeabilities and capillary pressure should be the functions of two (at least) variables (s, η). In other words, the parameters of two-phase flow should be different for various phase structures. Moreover, it is natural to expect that the mathematical type of the flow model may be different for various phase structures, not only the model parameters.

In this situation, the question which follows directly from this short analysis is about to what extent the classic Buckley–Leverett model may be used to describe the two-phase displacement in porous media, the process which has a high importance in petroleum engineering and groundwater theory. Indeed, the classic model is based on the assumption of a capillary equilibrium, which can really be reached only after the displacement is already finished and each phase flows inside its own channel sub-system. At the same time, the displacement process, characterized by a meniscus structure, is highly non-stationary, with a fast and high variation of the local saturation in time, so it is far away from any equilibrium state.

In the present paper we develop a new entirely closed model of two-phase flow through porous media, developed especially for the meniscus structure. This model is qualitatively different from all the other models suggested for two-phase flows. The basic difference is in the description of the interface movement and capillary phenomena.

The comprehension of the fact that the classic model is probably not sufficient to describe fluid displacements in porous media is manifested in a number of attempts to develop other models, based on other principles. Among these attempts we distinguish three main approaches: (i) introduction of phase interaction across the interface in the form of cross-terms (cross-term model); (ii) introduction of non-equilibrium ("dynamic") effects (dynamic models); (iii) introduction of the fluid–fluid interface as an independent third phase (model with an interface transport).

1.2. MODEL WITH CROSS TERMS

The model with cross terms was initially suggested independently in Kurbanov (1968), Raats and Klute (1968) and Shvidler (1961). According to this model, the phase velocity is proportional both to the pressure gradient in this phase and to the pressure gradient in the second phase, which reflects an interaction between the phases across the interfaces. Later, this model was developed within a phenomenological approach (Marle, 1982, Marinbah and Lusternik, 1985, Thigpen and Berryman, 1985, Barci *et al.*, 1990, Kalaydjian, 1990). In Whitaker (1986) and Quintard and Whitaker (1988) this model was obtained using the volume averaging method applied to the Stokes two-phase flow. A similar model was derived from the Stokes equations, using the asymptotic homogenization method in Auriault and Sanchez-Palencia (1986). In Whitaker (1986), Quintard and Whitaker (1988) and Auriault and Sanchez-Palencia (1986) the film phase structure was accepted: each phase was continuous, both phases were present in each porous channel. The form of the cross relative permeabilities was obtained by numerical simulations of the film structure in a single pore in Kalaydjian (1990), by a laboratory experiment in Kalaydjian *et al.* (1989), Zarcone and Lenormand (1994) and Rose (1989), and by the lattice-gas microscale simulations of Stokes flow in a representative elementary volume Rothman (1989).

The cross-term model reflects such a viscous interaction between two stratified fluids, when one fluid moving tangentially to the interface curries the other fluid into the flow. Therefore, this model can not be directly applied to describe the meniscus flow structure which is characterized by a frontal displacement of a fluid by another.

1.3. NON-EQUILIBRIUM (DYNAMIC) MODEL

Another approach to modify the flow model was developed in Barenblatt (1971), Barenblatt and Vinichenko (1980) and Vinichenko (1978). Assuming that the displacement is a non-equilibrium process, the authors introduced a non-equilibrium in the form of relative permeabilities as the functions of saturation and the time derivative of saturation. The non-equilibrium behaviour of phase permeabilities (a long-term stabilization) was shown in Entov and Chen-Sin (1987), Singal and Somerton (1977) and Blunt and King (1991), using numerical simulations in network models of porous media and experimental data.

The mechanisms which cause a non-equilibrium are various, therefore a number of different model versions exists. In Nikolaevskii *et al.* (1968), the non-equilibrium model was developed in terms of a phenomenological approach.

A physical mechanism to obtain this model was suggested in Barenblatt and Entov (1972) or Zarubin (1993). It is based on assuming the existence of two components of each phase: an active and a passive component. The exchange between them leads to a non-equilibrium behaviour. A differential exchange mechanism was suggested in Buevich and Mambetov (1991), where each phase was considered as a system of clusters of various scales. The exchange between every two clusters produces a non-equilibrium behaviour.

Another mechanism was illustrated in Bourgeat and Panfilov (1998) where the non-equilibrium is caused by medium heterogeneity. Between various components of a heterogeneous medium the non-instantaneous exchange takes place. The closed model with capillary relaxation was developed theoretically using an asymptotic homogenization procedure for two-phase flows in a heterogeneous medium.

The basic problem of this model is to construct the closure relation for a characteristic relaxation time. In Bocharov (1991) the relaxation time was estimated using experimental data; in Bourgeat and Panfilov (1998) an explicit analytical relation was obtained for the relaxation time.

The new interest in this kind of models was arisen in papers (Hasanizadeh and Gray, 1993b; Hasanizadeh, 1997).

The dynamic models, describing a relaxation of flow parameters, may thus be considered as a low deviation from the classic equilibrium model. This means that the interior phase structure represents a weak deviation from the classic channel structure which does not correspond to a meniscus structure.

1.4. MODEL WITH THE TRANSPORT OF INTERFACES

In the paper (Marle, 1982) where a system with a number of interfaces was examined within the framework of a phenomenological approach, a new idea was suggested to examine the highly dispersed interface as an individual continuum while formulating the mass and momentum balance equations for it. In a particular case, the cross-term model was obtained as a justification of the developed approach. Similar ideas but based on the thermodynamics methods of the mixture theory were developed in (Hasanizadeh and Gray, 1990, 1993a, b). The basic results of this approach was: (i) the insufficiency of the saturation as a unique function to describe the two-phase flow structure; (ii) the specific interface appearing as a new variable determining the two-phase system behaviour (the configuration parameter); (iii) a necessity to construct supplementary equations defining the interface movement; (iv) a possible modification of the momentum balance equation for each phase produced by the capillary forces. To simplify and to close the generic equations obtained for the interface movement the authors used the hypothesis of a weak non-equilibrium behaviour, which provided a possibility to use a simple kinetic equation. Thus, the interface behaviour is examined as a weak perturbation with respect to the equilibrium state. In the final form the model is qualitatively similar to the dynamic models examined in the previous paragraph.

The results of all these works opened new perspectives in the theory of two-phase flows. The concept of the specific interface as a new flow function was a sufficient advancement of the theory. At the same time, a weakly non-equilibrium description may be applied to such a highly non-equilibrium process as fluid displacement only as a first approximation, which is not much better than the classic equilibrium model.

1.5. SUGGESTED APPROACH

In the present paper the meniscus structure is examined independently of weakly non-equilibrium approaches.

In Section 2 we will show that the two-phase system in the vicinity of a meniscus behaves in a highly different manner than usually accepted. First of all, the flow velocities of both phases are identical near the meniscus, regardless of the individual mobility of each phase. The two-phase system behaves here as a unique continuum. Second, the movement of each meniscus is highly controlled by the capillary forces applied to it: this force can accelerate or decelerate the meniscus movement, thus playing the role of a vectorial function (a true force) determining the momentum transfer to the meniscus.

According to these observations, we suggest two new elements as a basis of the two-phase model. First, we introduce the concept of a meniscus continuum (M-continuum) as a third fluid phase (along with two traditional continuums associated to each individual fluid) which has some particular properties influenced by the individual properties of each fluid and by the presence of a system of rigid meniscus between them. Second, we introduce the concept of a vectorial field of capillary forces which enter, due to their vectorial origin, in the momentum balance equation. These two ideas are sufficient to develop a constructive phenomenological two-phase flow model. The closure relations are obtained using the results of numerical capillary network simulations.

The concept of the M-continuum is an advancement of the concept of a supplementary continuum presented by the interface suggested in Marle (1982) and Hasanizadeh and Gray (1990). The M-continuum is not simply an interface dissipated in space, it is a more complicated object including a part of both fluids with the interface and possessing some combined physical properties.

The concept of the vector capillarity is a new approach to describe the highly non-equilibrium capillary processes associated to the meniscus phase structure. According to this concept, a highly non-equilibrium capillarity has another vectorial origin (the classic capillarity is scalar). Such a radical change in the qualitative behaviour can not be evidently captured within a weakly non-equilibrium thermodynamic approach.

The first attempts to develop the meniscus model was published in Panfilova et al. (1998) and Panfilov and Panfilov (2000).

2. Properties of a Repetitive Meniscus Field

The following basic properties of the meniscus configuration may be revealed before constructing a mathematical model: (i) the law of meniscus passage through a pore, (ii) the roles of meniscus passage across a vertex, (iii) the statistical law of meniscus birth/death (compensation mechanism), (iv) a mechanism ensuring the formation of a spatially repetitive meniscus field.

2.1. PASSAGE OF A MENISCUS THROUGH A PORE

Let us examine the piston displacement of a fluid by another in a capillary channel, as shown in Figure 1.

Let us assume that the interface between the fluids called a meniscus has the following properties: (i) it crosses the pore wall along a closed curve which is a triple-contact curve, (ii) it is non-deformable, and (iii) it is plane but produces a Laplace capillary pressure calculated for a spherical interface. Within the framework of these assumptions, it is hard to apply the Navier–Stokes equations to describe the flow. Indeed, due to the non-slip conditions the triple-contact curve can move only if the creeping mechanism is triggered. However this mechanism leads to the interface deformations. In other words, the accepted assumptions are not strictly compatible with the microscale flow pattern. At the same time, they are correct on average, i.e. they are compatible with the Navier–Stokes equations integrated over the channel cross-section. When the cross-section is assumed to be uniform, or when the Reynolds number is assumed to be low, the mentioned averaging leads to the Poiseuille law. It can then be shown that the system of Poiseuille equations for each fluid does not contradict the assumptions mentioned above.

The system of Poiseuille equations (1a,b) with the interface (1c), exterior boundary-value and initial conditions (1d) takes the following form:

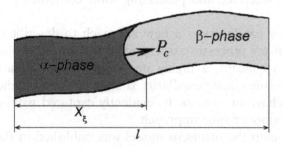

Figure 1. Two-phase displacement in a capillary.

$$-U_\alpha = \frac{\gamma r^2}{\mu_\alpha} \frac{dP_\alpha}{dx}, \quad 0 < x < x_\xi, \tag{1a}$$

$$-U_\beta = \frac{\gamma r^2}{\mu_\beta} \frac{dP_\beta}{dx}, \quad x_\xi < x < l, \tag{1b}$$

when $x = x_\xi$,

$$U_\alpha = U_\beta \equiv U, \quad P_\alpha|_{x=x_\xi - 0} = P_\beta|_{x=x_\xi + 0} - P_c, \quad \frac{dx_\xi}{dx} = U, \tag{1c}$$

$$P_\alpha|_{x=0} = P_A, \quad P_\beta|_{x=l} = P_B, \quad x_\xi|_{t=0} = 0, \tag{1d}$$

$$P_c(r) \equiv \frac{2\sigma \cos\theta}{r},$$

where l is the channel length, γ is a channel shape parameter (1/8 for a circular cross-section), r is the channel radius, x_ξ is the local coordinate of the mobile meniscus, P is the pressure, μ is dynamic viscosity, U is the flow velocity averaged over the channel cross-section, t is the local time calculated from the instant of the meniscus appearance at the inlet section, x is the local coordinate calculated from the inlet section, and σ, θ and P_c are the surface tension, the contact angle and the Laplace capillary pressure at the meniscus. The conditions at the interface represent the flow rate conservation, a pressure jump due to the capillary effect and the fact that the meniscus velocity is equal to the physical fluid velocity.

Mathematically this problem is correctly formulated and has a unique solution. Indeed, by eliminating consecutively the velocity and the pressure, we obtain a non-linear differential equation with respect to the meniscus coordinate, known as the Washburn equation

$$\frac{dx_\xi}{dt} = \frac{\gamma r^2 (\delta P + P_c(r))}{\mu_\alpha x_\xi + \mu_\beta (l - x_\xi)}, \quad x_\xi(0) = 0; \quad \delta P \equiv P_A - P_B, \tag{2}$$

which has a unique differentiable solution.

Let us rewrite this equation differently, by introducing $\xi = x_\xi / l$

$$U = \frac{\gamma r^2}{\langle \mu \rangle} \frac{(\delta P + P_c(r))}{l}, \tag{3}$$

where $\langle \mu \rangle = \mu_\alpha \xi + \mu_\beta (1 - \xi)$ is the averaged fluid viscosity in a pore. By introducing the volume saturation of the displacing phase, we note that

$$\langle \mu \rangle = \mu_\alpha s + \mu_\beta (1 - s). \tag{4}$$

Equation (3) shows two particular properties of the two-pase system behaviour in the vicinity of a meniscus. First of all, the system behaves as a united pseudo-phase advancing with a single velocity common for both true phases and with a viscosity of a weighted mean value between two individual viscosities. Second, let us rewrite Equation (3) in the vectorial form

$$\frac{\langle \mu \rangle}{\gamma r^2} \vec{U} = -\left(\frac{\partial P}{\partial x} - \frac{P_c(r)}{l} \right) \vec{e}_x,$$ (5)

where $\vec{e}_x$ represents the unit vector along axis x.

This is evidently a version of the momentum balance equation saying that the friction forces are balanced by a sum of pressure forces and capillary forces, with a negligible inertia force. Thus, the capillarity becomes a vectorial concept and enters into the momentum equation.

2.2. PASSAGE OF A MENISCUS THROUGH A VERTEX. COMPENSATION MECHANISM

The passage of a meniscus through a branching point of pore channels is highly different from the continuous motion in a regular channel. First of all, the birth of new mobile menisci is observed in a vertex, as well as the probable immobilization ("a death") of some menisci due to the capillary trapping mechanism. Second, the passage through a vertex constitutes an instantaneous jump, resulting from the surface instability. The variety of these phenomena observed on a random meniscus set is governed by some general statistical laws which enable us to describe its behaviour in an effective way.

According to the previous remarks, we assume that the time of meniscus jump through a vertex is much lower than the time of a slow viscous flow along a pore, so that the passage of a meniscus through a vertex may be considered as instantaneous. Thus, a vertex only plays the role of a meniscus source, which may be positive (a meniscus birth) or negative (a meniscus death).

We acknowledge next the meniscus conservation law: the average number of the mobile meniscus is conserved, so that the birth of new menisci is compensated, on average, by the death of other menisci. This conservation is ensured by the *meniscus compensation mechanism*. Let us examine an elementary network, called "the doublet" (Rose, 1967), which represents two channels of different radii connected between them in vertexes located periodically as shown in Figure 2.

The displacing phase (grey) enters into the network via the left-hand cross-section and forms two mobile menisci (1 and 2 in Figure 2(a)). Meniscus 1 reaches the first vertex more rapidly (Figure 2(b)), crosses it and forms two new menisci, 3 and 4. At the same time, meniscus 2 and 4 are immobilized by the trapping of the displaced phase between sections 2 and 4. Thus, the birth of the new meniscus 3 will be compensated by the death of meniscus 2. As a result, after a vertex is passed, the number

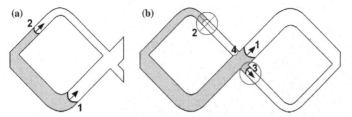

Figure 2. Illustration to the meniscus compensation mechanism.

of mobile menisci is conserved. These two menisci, 1 and 3, continue their advancement. In the next vertex, similar events will be observed.

The compensation mechanism completed by assuming an instantaneous passage through each vertex enables us to neglect the vertex impact on the two-phase flow on average.

2.3. A REPETITIVE MENISCUS FIELD: DUPLICATION MECHANISM

The basic assumption of the present paper says that the mobile menisci form a repetitive space field. This assumption enables us to examine the meniscus field on average. However, the formation of such a field is not evident. In particular, in a simple doublet model, used in the precedent section, no mobile menisci exist behind two frontal menisci. However, numerical simulations in the network capillary models show that this assumption is true when the medium is heterogeneous. Figure 3 illustrates a typical fluid repartition within the porous space for a displacement process. This image is obtained as a result of numerical simulations based on the network model of porous media, which is described both in Section 4.2 and in Panfilova (2003) and Panfilova and Muller (1996).

The displacing phase enters via the top section; the lateral vertical sections are impermeable; the square network is made of cylindrical tubes of random radius; the displacement in a tube is piston-like; the flow of each phase in each tube is governed by the Poiseuille law. The displaced phase cannot be displaced if this phase is not connected to the outlet section, or when stopped by a capillary counter-pressure. The inlet and outlet pressures are maintained invariable in time.

As seen, the mobile menisci form a statistical highly dispersed repetitive field uniformly covering all the space.

The hypothesis on a repetitive meniscus field may be clearly justified when using classical methods of analysis suited to heterogeneous media. In particular, this can be done by using the effective medium method.

Let us examine the model of porous medium formed by an individual porous channel connected to the effective medium, as shown in Figure 4. The general effective medium method is based on immersing a medium

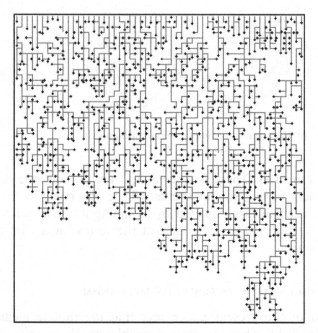

Figure 3. A typical phase structure in porous space for a displacement process: grey lines: the displacing phase; the black points are the mobile menisci.

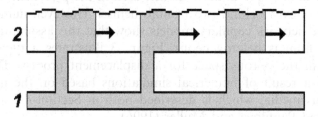

Figure 4. The duplication mechanism in an effective medium model for transport phenomena: 2 is the effective medium, 1 is a highly permeable individual porous channel.

inclusion into the effective medium. We examine a particular case of the method when the individual inclusion is a pore channel with all its connections. In this case, the effective medium is evidently averaged over the transverse sections, not over the volume.

In the individual channel, the motion of a true meniscus is analyzed, whilst in the effective medium one studies the average displacement front.

Let us establish the laws of meniscus propagation through such a model. Two different cases may be observed. First of all, let the flow velocity in the individual channel be lower than that in the effective medium. Once the average front in the effective medium reaches a vertex, the meniscus in the individual channel is blocked, thus forming a trapped phase. This mechanism is similar to that described in the doublet model of the preceding paragraph.

Let us now examine the second case, when the flow velocity in the individual channel is greater than in the effective medium. Due to this, we will call such an individual channel "percolating". Once the meniscus in the individual channel reaches a vertex, it penetrates into the effective medium and gives rise to a new average front. The new front does not block the precedent front as in an effective medium, the displaced phase can always find a way to leave the medium. Thus, each vertex crossed by the true meniscus in the individual channel gives rise to new mobile fronts, by forming a repetitive mobile meniscus field, as seen in Figure 4.

As shown, the repetitive meniscus field is formed if the pore radii are not uniform and if at least one individual channel more penetrable than the average channel exists. The total structure of a mobile meniscus field is presented in Figure 5 which shows an average channel with branches ensuring the duplication mechanism. The value h_c, which is the mean distance between two neighbouring mobile menisci, plays the role of a phase configuration parameter.

3. Phenomenological Approach to Construct the Meniscus Model

We assume that the total two-phase system in a porous medium represents a superposition of three crossed continua coexisting at each space point and at each time instant

(I) the *M-continuum* ("meniscus continuum") which represents a repetitive mobile meniscus field,

(II) the α-continuum which corresponds to the system of pore channels occupied by the displacing phase only, and

(III) the β-continuum which corresponds to the channels filled up by the displaced phase only.

A phenomenological approach ignores any interior phase configuration, so the true configuration formed by a repetitive meniscus field should be introduced into the properties of the M-continuum.

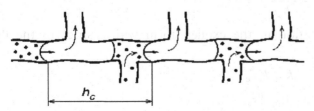

Figure 5. Total structure of a mobile meniscus field.

3.1. M-CONTINUUM

We admit that the M-continuum is formed by the repetitive system of mobile menisci together with the two fluid phases separated by these menisci.

Based on the properties of the repetitive meniscus field described in the previous section, let us introduce the notion of the M-continuum.

1. At each space point and each time instant, the M-continuum has a unique flow velocity $U=U_\alpha=U_\beta$. This means that, within the M-continuum, both phases move as a unit ensemble.
2. The viscosity of the fluid in the M-continuum is $\mu=\mu_\alpha s^I+\mu_\beta(1-s^I)$ with s^I as the local saturation of the α-phase within the M-continuum.
3. A continuous vector field of capillary forces is distributed over the M-continuum with the density per unit length

$$\frac{P_c \vec{e}_c}{h_c}, \qquad P_c \equiv \frac{2\sigma \cos\theta}{\langle r \rangle} \tag{6}$$

where $\vec{e}_c$ is the local unit vector of this field, P_c is the Laplace capillary pressure applied to an average meniscus, $\langle r \rangle$ is the mean pore radius, σ and θ are the surface tension and the contact angle.

4. The unit capillary vector $\vec{e}_c$ is parallel to the flow velocity of the M-continuum and is directed from the wetting phase towards the non-wetting one.

Note that the equivalence of phase velocities takes place only within the M-continuum, i.e. in the vicinity of menisci. At the same time, the total phase velocities defined as $\langle U_\alpha \rangle = U^M \alpha^M + U^\alpha \alpha^\alpha$ and $\langle U_\beta \rangle = U^M \alpha^M + U^\beta \alpha^\beta$ are not identical.

The mentioned properties are necessary in order to obtain the momentum balance equation similar to (3). Indeed, an elementary volume of the M-continuum undergoes the action of three kinds of forces: the pressure force imposed by the environmental fluid; the friction force which is proportional to the flow velocity and continuum viscosity and the distributed capillary force defined above. Then the momentum balance equation takes the form

$$-\text{grad } P + \frac{\langle \mu \rangle \vec{U}}{K_M} - \frac{P_c \vec{e}_c}{h_c} = 0. \tag{7}$$

Rewriting this equation in the explicit form with respect to the flow velocity, we obtain

$$\vec{U} = -\frac{K_M}{\langle \mu \rangle}\text{grad } P + \frac{K_M}{\langle \mu \rangle}\frac{P_c \vec{e}_c}{h_c}, \tag{8}$$

which is similar to (3).

3.2. PHENOMENOLOGICAL FLOW EQUATIONS

We can now formulate the total system of flow equations for three crossed continua. We examine only the isothermal processes when the energy equation may be ignored. Let the superscripts I, II and III be used to indicate the M-continuum, the α-continuum and the β-continuum correspondingly. Let K^i and ϕ^i be the intrinsic permeability and porosity of the continuum i, α^i be the volume fraction of the continuum i, s^I be the internal saturation of the displacing phase in the M-continuum. The intern saturation is 1 for continuum I and 0 for continuum II. Let us use the Darcy velocity $\vec{V}$ instead of the true velocity $\vec{U}$, using the following relation: $\vec{V}^i = \vec{U}^i \phi^i$, $(i = I, II, III)$. This system of mass conservation and momentum balance equations for three continua is

$$\phi^i \frac{\partial \left(\rho^i \alpha^i\right)}{\partial t} + \mathrm{div}\left(\rho^i \alpha^i U^i\right) = 0, \quad i = I, II, III, \tag{9}$$

$$\vec{V}^I = \frac{K^I}{\mu^I}\left(-\mathrm{grad}\ P^I + \frac{P_c}{l}\eta_c \vec{e}_c\right), \tag{10}$$

$$\vec{V}^{II} = -\frac{K^{II}}{\mu_\alpha}\mathrm{grad}\ P^{II}, \tag{11}$$

$$\vec{V}^{III} = -\frac{K^{III}}{\mu_\beta}\mathrm{grad}\ P^{III}, \tag{12}$$

$$\alpha^I + \alpha^{II} + \alpha^{III} = 1, \tag{13}$$

where $\rho^I = \rho_\alpha s^I + \rho_\beta(1 - s^I)$ et $\mu^I = \mu_\alpha s^I + \mu_\beta(1 - s^I)$ are the density and viscosity of the M-continuum; $\rho^{II} = \rho_\alpha$, $\rho^{III} = \rho_\beta$, l is a mean pore length.

The meniscus concentration, η_c, is introduced instead of the distance between menisci, h_c

$$\eta_c \equiv \frac{l}{h_c}.$$

Note that η_c is the intern concentration within the M-continuum.

This value is consistent to the definition of a concentration, which varies between 0 and 1. The value 0 is reached when the distance between menisci is much larger than the pore length, whereas 1 corresponds to the minimal admissible distance between two neighbouring menisci, which is equal to l (a pore may contain no more than one mobile meniscus).

We also accept the condition of a uniform phase distribution within the pore space

$$K^I = K^{II} = K^{III} \equiv K, \quad \phi^I = \phi^{II} = \phi^{III} \equiv \phi \tag{14}$$

and the condition of an equilibrium between continua

$$P^I = P^{II} = P^{III} \equiv P. \tag{15}$$

We also accept that both phases are incompressible.

The systems (9)–(15) may be transformed into the following system:

$$\phi\frac{\partial\left(\rho^I\alpha^I\right)}{\partial t}=\mathrm{div}\left(\frac{K\rho^I\alpha^I}{\mu^I}\left(\mathrm{grad}\,P-\frac{P_\mathrm{c}}{l}\eta_\mathrm{c}\vec{e}_\mathrm{c}\right)\right),\tag{16}$$

$$\phi\frac{\partial\alpha^{II}}{\partial t}=\mathrm{div}\left(\frac{K\alpha^{II}}{\mu_\alpha}\mathrm{grad}\,P\right),\tag{17}$$

$$\phi\frac{\partial\alpha^{III}}{\partial t}=\mathrm{div}\left(\frac{K\alpha^{III}}{\mu_\beta}\mathrm{grad}\,P\right),\tag{18}$$

$$\alpha^I+\alpha^{II}+\alpha^{III}=1.\tag{19}$$

This system contains four equations and six variables: P, $\{\alpha^i\}_{i=1}^3$, s^I (or ρ^I), and η_c.

To close this system, two supplementary relations must be added.

4. Closure Relations

The definition of two variables, s^I and η_c, which are responsible for the internal structure of the M-continuum, can not be determined within the framework of the phenomenological approach. They are thus free parameters of the model. To determine them, another finer approach should be applied, based either on a theoretical quantitative analysis of the micro-mechanisms of the repetitive meniscus configuration, or on the statistical treatment of data resulting from microscale numerical simulations.

4.1. THEORETICAL MODEL OF A STOCHASTIC MENISCUS STREAM

Let s be the total α-saturation. When the system becomes single-phase ($s=0$ or $s=1$), then the meniscus concentration is zero. For a rather uniform presence of both phases, the meniscus concentration is maximal. Thus, the meniscus concentration seems to be correlated with the saturation. In the present section we develop the model of a stochastic meniscus stream which yields a strict theoretical relationship between η_c and s which will be justified next by numerical simulations.

Let us introduce the total meniscus concentration in the overall medium: $\eta\equiv\alpha^I\eta_\mathrm{c}$, instead of η_c. Let us examine a fragment of the capillary network between two neighbouring vertexes (Figure 6(a)).

It represents a bundle of N parallel capillaries of a same length l equal to the mean size of a pore (Figure 6b). The mean distance between two adjacent capillaries in the examined fragment is also l. The capillary radii are random. Each capillary is filled up by β-phase at the initial state. Let a family of meniscus approach to the inlet section of the examined medium fragment and enters into it during the period $(0, T_\mathrm{ap})$. Let t_ap be

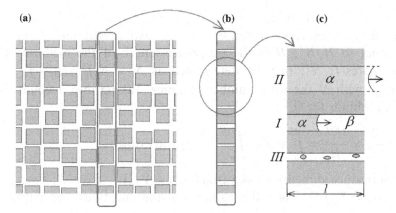

Figure 6. Theoretical model of a random meniscus appearance in a fragment of the capillary network.

the random time of meniscus appearance in the segment, V be the meniscus velocity. The time origin corresponds to the instant of the appearance of the first meniscus. We assume that a fixed part of capillaries when the velocity is below a critical value V_0 is not accessible to meniscus and is always filled up with β-phase. To introduce the residual α-saturation, we assume that these non-accessible pores contain also a trapped α-phase in a dispersed form. The volume fraction of the dispersed α-phase in non-accessible pores is ω. The total trapped α-saturation is s_*. Let $f(V)$ and $\psi(t_{ap})$ be the probability density of the meniscus velocity and the appearance time normalized such that

$$\int_0^\infty f(V)dV = 1, \qquad \int_0^\infty \psi(t_{ap})dt_{ap} = 1.$$

For a fixed time instant t, all the pores form three classes shown in (Figure 6(c))

- the class III where $V \leqslant V_0$ will contain β-phase and trapped α-phase,
- the class I where $V_0 < V$ and $V(t - t_{ap}) \leqslant l$ will contain mobile menisci,
- the class II where $V(t - t_{ap}) > l$ will be filled up with α-phase.

The volume of mobile α-phase is l in a pore II and $V(t - t_{ap})$ in a pore I, if a meniscus entered into this pore during the period $(0, t)$. Then we obtain for the α-phase saturation

$$s(t) = \frac{1}{l} \int_0^t \psi(t_{ap}) \int_{V_0}^{l/(t - t_{ap})} V \cdot (t - t_{ap}) f(V)dV dt_{ap}$$

$$+ \int_0^t \psi(t_{ap}) \int_{l/(t - t_{ap})}^\infty f(V)dV dt_{ap} + s_*.$$

The meniscus total concentration $\eta \equiv \eta_c \alpha^I$ is $l/\langle h \rangle$ by definition, where $\langle h \rangle$ is the mean transverse distance between two neighbouring menisci which is equal to H/N^I; H is the transverse length of the medium. The number of pores-I, N^I, may be determined as $N^I = N \int_0^t \psi(t_{ap}) \left(\int_{V_0}^{l/(t-t_{ap})} f(V) dV \right) dt_{ap}$, with N the total number of pores. Taking into account that $H/N = l$, we obtain for η

$$\eta(t) = \frac{l}{\langle h \rangle} = \frac{lN^I}{H} = \int_0^t \psi(t_{ap}) \int_{V_0}^{l/(t-t_{ap})} f(V) dV dt_{ap}.$$

Let us introduce dimensionless variables: $v = V/\langle V \rangle$, $\tau = t/T$, $\theta = t_{ap}/T$, $T = l/\langle V \rangle$, $\tau_* = T_{ap}/T$. Then the saturation and meniscus concentration become

$$s(\tau) = \int_0^\tau \psi(\theta)(\tau - \theta) \int_{v_0}^{1/(\tau-\theta)} vf(v) dv d\theta + \int_0^\tau \psi(\theta) \int_{1/(\tau-\theta)}^\infty f(v) dv d\theta + s_*,$$

$$\eta(\tau) = \int_0^\tau \psi(\theta) \int_{v_0}^{1/(\tau-\theta)} f(v) dv d\theta.$$

These two relationships determine parametrically the function $\eta(s)$.

For a uniform distribution of the appearance times and an asymmetric Rayleigh distribution of the velocities

$$\psi(\theta) = \begin{cases} \dfrac{1}{\tau_*}, & 0 < \theta \leqslant \tau_*, \\ 0, & \theta > \tau_*, \end{cases} \qquad f(v) = 2\xi v e^{-\xi v^2}, \quad v > 0, \quad \xi = \frac{\pi}{4},$$

the results of calculations are shown in Figure 7(a) for three various parameters v_0 (the lower curve corresponds to a greater value of v_0). Note that parameter v_0 determining that part of pores which is non-accessible for the invading fluid is inversely proportional to the capillary number (the ratio of the pressure difference to the mean capillary pressure).

If $\psi(\theta)$ is the delta-function, then it is possible to obtain an analytical relation for $\eta(s)$ in a parametric form

$$\eta(\tau) = \omega \left\{ e^{-\xi v_0^2} - e^{-\xi/\tau^2} \right\}, \tag{20a}$$

$$s(\tau) = \tau e^{-\xi v_0^2} + \frac{\tau}{2} \sqrt{\frac{\pi}{\xi}} \left[\text{erf} \left(\frac{\sqrt{\xi}}{\tau} \right) - \text{erf} \left(\sqrt{\xi} \, v_0 \right) \right] + s_*, \tag{20b}$$

where ω is a new parameter introduced in order to fit the data of numerical simulations (see the next section).

According to the results obtained we conclude that: (i) the variable η may be examined as the function of s; (ii) the function $\eta(s)$ is always non-monotonic and asymmetric; (iii) this function may depends parametrically on the capillary number; (iv) the qualitative behaviour of this function may be described by relations (20).

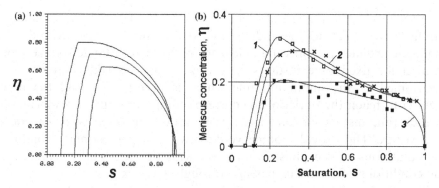

Figure 7. Total meniscus concentration versus saturation: (a) – theoretical meniscus stream model; (b) – results of capillary network simulations: $Ca = 10^2$ (1), $Ca = 1$ (2), and $Ca = 10^{-3}$ (3).

4.2. CLOSURE RELATIONS FOR $\eta(s)$ FROM CAPILLARY NETWORK SIMULATIONS

To justify the accepted hypothesis concerning the meniscus concentration we performed a series of microscale numerical simulations based on the capillary network model of porous media. The corresponding numerical algorithm, developed by the authours, is described in various preceding publications (Panfilova and Muller, 1996; Panfilov, 2000; Panfilova and Panfilov, 2000).

Shortly, the geometrical model of porous medium represents a uniform cubic network of vertexes connected by cylindrical capillary tubes. The tube radius is considered as a stochastic stationary space function with given global distribution, correlation function and mean value. The two-phase distribution structure in porous space is of meniscus type, so if two phases are present in a pore tube, the interface between them has a form of a meniscus. The flow of each phase in each pore tube is described by the Poiseuille law with a pressure discontinuity at each meniscus (1). The flow conservation applied to each vertex yields a system of recurrent equations formulated with respect to the vertex pressures. As this system may be considered as a discrete form of a differential Laplace-like equation, the solution to this system is constructed by using standard methods of solving the elliptic equations.

The passage through a vertex is instantaneous when permitted. A meniscus may pass through a vertex into a tube connected to this vertex, if the capillary counter-pressure in this tube is not sufficient to stop it.

A supplementary condition limiting motion is imposed by the phase connectivity. A part of the displacing phase is assumed to be immobile if this part has no connection to the inlet section. A part of the displaced phase is immobile if this part has no connection to the outlet section.

As a result, the numerical algorithm consists of two basic modules: (i) the hydrodynamic module and (ii) the logical module. The hydrodynamic module calculates the pressure field and the local coordinates of the mobile menisci for fixed time instants. The time step is variable in time and is determined as the minimum time which is necessary for all the mobile menisci to reach the outlet of the corresponding pore tube. This means that at each time instant we have exactly one mobile meniscus which is located in a vertex. The logical module verifies two kinds of mobility conditions: the conditions of the phase connection with the inlet/outlet sections, and the conditions of meniscus passage through a vertex.

To calculate the total meniscus concentration, $\eta(s)$, we have used the numerical data obtained for various sections orthogonal to the macroscopic flow direction. In each section we have calculated the mean meniscus concentration (inversely proportional to the mean distance between the neighbouring menisci) and the saturation. Thus, we have obtained the pair of points $\{s, \eta\}$. The results of simulations are shown in Figure 7b for a fixed viscosity ratio (0.5) and three different values of Ca.

The non-monotonous form of this function, predicted by the theoretical analysis, is weakly variable and, thus, may be considered as independent of parameters Ca and $\bar{\mu}$. All these curves may be approximated by relation (20) with $v_0 = 0.5$, $s_* = 0.16$, $\omega = 0.4$.

4.3. CLOSURE RELATION FOR THE INTERNAL SATURATION OF THE M-CONTINUUM s^I

Saturation s^I can be obtained by examining the evolution of the saturation field in the effective medium model (Figure 4). In this paper, we only suggest the outline of such a qualitative analysis, which is sufficient to propose a simpler closure relation.

The filling of each cell of the effective medium by the displacing phase coming from the percolating channel (Figure 4), starts from the inlet cell and then continues, with a delay, in the consecutive cells. Therefore, the cell saturation varies in space and in time. Nevertheless, if we assume that the filling velocity of the effective medium is much lower than the velocity of medium percolation through the percolating channel, the delay and the difference between the cells will be removed. Within this approximation, the cell saturation is independent of space coordinates.

In an arbitrary cell, the saturation grows in time but up to that instant only when the mobile front (meniscus) meets the immobile front which marks the inlet section of any new cell. After this, the cell saturation remains invariable and equal to the constant value s_*. The filling time, $t_{fil}\ h_c$ is much lower than the macroscopic time (L, with L the macroscale), if $h_c \ll L$ (a high meniscus concentration). Hence, it is admissible to ignore the details of the true saturation variation during these short periods and to accept that

$$s^I \approx \text{const} = s_*. \tag{21}$$

The value s_* can be estimated as $s_* = 1 - s_\beta^{\text{trap}}$ where s_β^{trap} is the classical residual saturation of the displaced phase.

It is necessary to note that the volume of the displacing phase in the M-continuum should disappear when the total saturation tends to zero. This does not contradict relation (21) because the volume of the displacing phase in the M-continuum will decrease due to the decrease of the total volume of the M-continuum it-self.

5. Closed Meniscus Model

5.1. MODEL FORMULATED IN TERMS OF THE TOTAL SATURATION

First of all, the density ρ^I becomes constant due to (21) and then may be removed from Equation (16). Using closure relations (21) and (20), we can now transform the models (16)–(19). Let us introduce the total saturation of the displacing phase, s

$$s = \alpha^I s^I + \alpha^{II}. \tag{22}$$

Multiplying Equation (16) by s^I (which is constant due to (21)) and adding Equations (16) and (17), we obtain the equation for the total saturation s (23). Adding all three Equations (16)–(18), we obtain Equation (24). The third equation of the system remains invariable (we take Equation (17)). The new system is

$$\phi \frac{\partial s}{\partial t} = \text{div}\left(\lambda_\alpha \text{grad } P - V_c s^I \eta(s) \vec{e}_c\right), \tag{23}$$

$$\text{div}(\lambda \text{grad } P - V_c \eta(s) \vec{e}_c) = 0, \tag{24}$$

$$\phi \frac{\partial \alpha^{II}}{\partial t} = \text{div}\left(\frac{K\alpha^{II}}{\mu_\alpha} \text{grad } P\right), \tag{25}$$

where $V_c = \frac{KP_c}{\mu^I l}$ is the characteristic capillary flow velocity

$$\lambda = \lambda_\alpha + \lambda_\beta, \quad \lambda_\alpha = \frac{Kk_\alpha}{\mu_\alpha}, \quad \lambda_\beta = \frac{Kk_\beta}{\mu_\beta}, \quad \overline{\mu} = \frac{\mu_\alpha}{\mu_\beta}$$

with new relative permeabilities

$$k_\alpha \left(s, \alpha^{II}\right) = \frac{\left(s - \alpha^{II}\right) \mu_\alpha}{\mu^I} + \alpha^{II},$$

$$k_\beta \left(s, \alpha^{II}\right) = 1 - \alpha^{II} - \frac{\left(s - \alpha^{II}\right) \mu_\alpha}{\mu^I}, \tag{26}$$

$\mu^I = \mu_\alpha s^I + \mu_\beta (1 - s^I)$. Function $\eta(s)$ is defined as (20).

Note that the relative permeabilities depend also on two constant parameters, s^I and $\overline{\mu}$.

First Equation (23), may be transformed into a saturation transport equation

$$\phi\frac{\partial s}{\partial t} - (\lambda \text{ grad } P - V_c\eta(s)\vec{e}_c) \text{ grad } F + \text{div}(V_c F_c \vec{e}_c) = 0 \qquad (27)$$

with F and F_c the fractional flow and the capillary flux defined as

$$F(s, \alpha^{II}) = \frac{\lambda_\alpha}{\lambda} = \frac{k_\alpha}{k_\alpha + k_\beta\overline{\mu}}, \qquad F_c(s, \alpha^{II}) = (s^I - F)\eta(s). \qquad (28)$$

To obtain this relation, let us note that

$$\text{div}(\lambda_\alpha \text{grad } P) = \text{div}(\lambda F \text{ grad } P) = F\text{div}(\lambda \text{ grad } P) + \lambda \text{ grad } P \text{ grad } F.$$

Then, using (24)

$$\text{div}(\lambda_\alpha \text{ grad } P) = F\text{div}(V_c\eta(s)\vec{e}_c) + \lambda \text{ grad } P \text{ grad } F.$$
$$= \text{div}(V_c F\eta(s)\vec{e}_c) + (\lambda \text{ grad } P - V_c\eta(s)\vec{e}_c) \text{ grad } F,$$

which leads directly to (27).

5.2. ASYMPTOTIC MODEL OF RELATIVE PERMEABILITIES

Systems (23)–(25) might be sufficiently simplified, if we accept the asymptotic model for the relative permeabilities, by assuming that the volume of the M-continuum is small. In this case $s^I \to 0$, then $\alpha^{II} \to s$, which follows from (22). Then the asymptotic relation for α^{II} may be presented in the following form:

$$\alpha^{II} = \alpha^{II}(s) = s^{\gamma+1}, \qquad \gamma = \gamma(s) \to 0, \quad \text{when } s \to 0. \qquad (29)$$

Let us determine parameter γ in such a way that relation (29) would be also used in the vicinity of the limit $s \to 1$, not only when $s \to 0$. The exponent γ must satisfy an evident condition: $\alpha^I + \alpha^{II} \leqslant 1$. According to (22): $\alpha^I = (s - \alpha^{II})/s^I$, thus using (29) we obtain the following inequality: $f(s) \equiv s - s^{\gamma+1}(1 - s^I) \leqslant s^I$. For two limit values of s, function $f(s)$ verifies this inequality: $f(0) = 0 \leqslant s^I$, $f(1) = s^I \leqslant s^I$. If the behaviour of $f(s)$ is monotonous, then the inequality is satisfied. Hence, it is sufficient to eliminate the maxima of this function.

For a derivative we have: $f' = 1 - (\gamma + 1)s^\gamma(1 - s^I)$. At peak: $s_{\max}^\gamma = 1/[(\gamma + 1)(1 - s^I)]$. The maximum between points 0 and 1 is not reached if $s_{\max} \geqslant 1$. This condition is satisfied if $0 \leqslant \gamma \leqslant s^I/(1 - s^I)$. In particular, a linear function $\gamma = (s^I/(1 - s^I))s$ satisfies these conditions.

Due to this approximation, the relative permeabilities become the functions of the saturation only, as well as of three parameters: s^I, $\bar{\mu}$ and γ.

This approximation replaces Equation (25) (which degenerates when the volume of the M-continuum is small) and reduces systems (23)–(25) to two Equations (23) and (24), or (27) and (24)

$$\phi\frac{\partial s}{\partial t} - (\lambda(s)\text{grad } P - V_c\eta(s)\vec{e}_c)\text{grad } F(s) + \text{div}(V_c F_c(s)\vec{e}_c) = 0, \qquad (30a)$$

$$\text{div}(\lambda(s)\text{grad } P - V_c\eta(s)\vec{e}_c) = 0. \qquad (30b)$$

5.3. SINGLE DIMENSION CASE

In the case of a single dimension flow, system (30) may be reduced to a single differential equation of saturation transport

$$\phi\frac{\partial s}{\partial t} + \frac{\partial}{\partial x}\left[V(t)F(s) + V_c e_c F_c(s)\right] = 0, \qquad (31)$$

where

$$e_c = \begin{cases} 1 & \text{if the displacing phase is wetting,} \\ -1 & \text{otherwise.} \end{cases} \qquad (32)$$

$V(t)$ is the total Darcy velocity of both phases, which appears as the result of an explicit integration of Equation (30b)

$$\lambda(s)\frac{\partial P}{\partial x} = V_c\eta(s)e_c - V(t). \qquad (33)$$

5.4. QUALITATIVE EFFECTS

Certain physical effects associated to the suggested model can be detected without solving any mathematical problem.

5.4.1. Hyperbolic Capillarity
Model (31) is a hyperbolic non-linear first-order equation. All its properties are determined by the structure of the sum of two functions: the non-capillary fractional flow $F(s)$ and the capillary flux $F_c(s)$. According to the physical meaning, function $F(s)$ is equal to the displacing phase flow rate divided to the total flow rate when the capillarity is nil. Function $F_c(s)$ is equal to the displacing phase flow rate caused only by capillary forces divided to the total flow rate.

Therefore, the capillarity entering in the model is in the form of an advection flux and keeps the hyperbolic type of the transport equation. This is qualitatively different from the classical Buckley–Leverett (or

Rapoport–Leas) model where the capillarity, appearing in the form of a diffusion flux, determines the parabolic type of the saturation transport equation.

5.4.2. *Viscous Deceleration (by the M-continuum)*

According to (26), the new relative permeabilities now depend on the viscosity ratio. This has a physical explanation. The mobility of the displacing phase is the M-continuum is determined by the mean viscosity $\mu^I = \mu_\alpha s^I + \mu_\beta (1 - s^I)$, whilst in the continuum I its mobility is controlled by its own viscosity μ_α. Let the displacing phase have no viscosity ($\mu_\alpha = 0$). Nevertheless, its integral mobility will be controlled by the viscosity of the M-continuum, which will be finite: $\mu^I = \mu_\beta (1 - s^I)$. So, the M-continuum decelerates the motion of the non-viscous displacing phase, which produces a decrease in the relative permeability k_α.

6. Analysis of the Model

6.1. FORMULATION OF A DISPLACEMENT PROBLEM

Let us examine a 1D problem of fluid displacement in a porous media, using model (31). In the initial state the medium is filled up with the phase β. The phase α is injected into the medium via the inlet section $x = 0$ with a permanent inlet saturation $s = 1$. The injection rate, Q, is known and constant in time; the lateral injection section is A, the medium length is L; the medium is homogeneous. For a qualitative analysis, let us neglect the residual (trapped) saturation of both phases. These conditions are sufficient to determine the flow velocity in (31): $V = Q/A$.

Let us introduce the dimensionless variables: x scaled using L; τ scaled using the time $t_* = L\phi/V$. Then the described displacement problem can be described by the following initial problem using model (31)

$$\frac{\partial s}{\partial \tau} + \frac{\partial G(s)}{\partial x} = 0, \quad G(s) \equiv F(s) + \frac{e_c}{Ca} F_c(s), \quad \tau > 0, \quad x \in R^1,$$
$$s|_\tau = 1 - H(x), \tag{34}$$

where $H(x)$ is the Heaviside function, $G(s)$ is the summary flux; functions $F(s)$ and $F_c(s)$ are defined using (28), (26), (29) and (32).

The solution is controlled by the structure of function $G(s)$, which depends on four parameters: s^I, $\overline{\mu}$, e_c and γ. For the analysis we have fixed two parameters: $s^I = 0.8$ and $\gamma = 4.9$. Using different values of other parameters, we obtain different versions of the problems which have qualitatively different solutions. Functions $F(s)$ and $F_c(s)$ are presented in Figure 8.

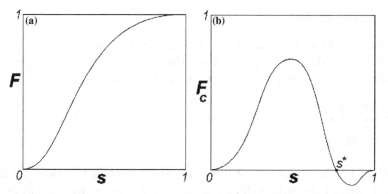

Figure 8. Fractional flow $F(s)$ and capillary flux $F_c(s)$.

6.2. SUPPLEMENTARY CONDITIONS FOR DISCONTINUOUS SOLUTIONS

As usual for hyperbolic systems, a problem similar to (34) may have discontinuous solutions only. A saturation discontinuity will be called "the front". Each front is characterized by its true transport velocity, U_f, the saturation on the upstream side of the front, s^-, and the saturation on the downstream side, s^+. These three values can be defined through three relations, which are the necessary and sufficient conditions of solution existence and uniqueness (and stability), (Rhee *et al.*, 1986)

(i) the Hugoniot mass balance condition, which is the integral analog of the differential equation in the vicinity of the front

$$U_f = \frac{G^+ - G^-}{s^+ - s^-}, \tag{35}$$

(ii) the Lax local evolutionary (or entropy) condition, Lax (1971)

$$G'(s^+) \leqslant U_f \leqslant G'(s^-), \tag{36}$$

which says that the front velocity cannot exceed the transport velocity of the saturation before the front and cannot be lower than the saturation behind the front.

(iii) the Oleinik global evolutionary condition, (Oleinik, 1959), which says that any tangent line $G'(s^-)$ should be located above any straight line relating points (s^-, G^-) and (s^+, G^+).

In the example considered here, the saturation transport velocity can be equal to the front velocity on only one side of the front.

6.3. COUNTER-CURRENT CAPILLARY IMBIBITION

Examine the problem (34) when $Ca \rightarrow 0$ and the total flow velocity is zero. Keeping the time derivative (otherwise, the equation changes type) and rescaling the time ($\tau \rightarrow \tau/Ca$), we obtain the following problem:

$$\frac{\partial s}{\partial \tau} + \frac{\partial F_c(s)}{\partial x} = 0, \qquad \tau > 0, \quad x \in R^1, \qquad (37)$$
$$s|_\tau = 1 - H(x).$$

According to the meaning of this equation, each saturation value is transported with transport velocity $F'(s)$, which is a non-monotonous function of saturation, as this results from Figure 8. In particular, the transport velocity is positive for small saturations and is negative for rather high saturation values. The formal solution to this equation, shown in Figure 9(b) (dashed line), is a non-unique function that has no physical meaning. Therefore, the problem has no continuous solutions. A physically admissible solution can be obtained only in the class of discontinuous functions.

It is possible to show that in the case of function $G = F_c$, the Hugoniot, Lax and Oleinik conditions determine two fronts

(i) a positive front is defined as:

$$s^+ = 0, \quad s^- = s_2, \quad U_f = F'_c(s_2),$$

where the value s_2 is the solution to a non-linear equation

$$F'_c(s_2) = \frac{F_c(s_2)}{s_2}, \qquad (38)$$

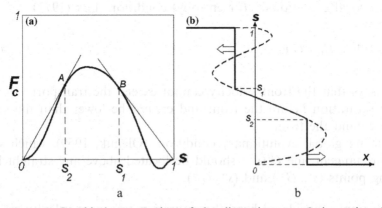

Figure 9. Graphical construction of the discontinuous solution using the diagram $F_c - s$ (a), and solution to the problem of capillary imbibition (b).

(ii) a negative front is defined as

$$s^+ = 1, \quad s^- = s_1, \quad U_f = F'_c(s_1),$$

where the value s_1 is the solution to the equation

$$F'_c(s_1) = -\frac{F_c(s_1)}{1 - s_1}. \tag{39}$$

The geometrical interpretation for these relations is shown in Figure 9a through the use of triangles OAS_2 and $1BS_1$.

For instance, for Equation (38), the right-hand part is equal to the ratio of two legs AS_2 and $0S_2$, whilst the left-hand part is equal to the tangent of the line OA. Thus, saturation s_2 is defined as the coordinate of the tangent point of the curve $F_c(s)$ by a tangent line coming from point $(0, 0)$.

In a similar way, saturation s_1 is defined by the tangent line coming from point $(1, 0)$.

The lines $0A$ and $1B$ are known as the Weldge tangents in the theory of oil recovery, or the Oleinik envelop is the theory of hyperbolic systems.

The discontinuous solution to the problem (35) is show in Figure 9(b) (solid curve).

It is seen that along with the usual front going into the medium, another second, "negative", front comes out of the medium with a negative velocity. Physically, this corresponds to the counter-current flow across the inlet section. Therefore, the suggested model is able to describe the counter-current imbibition, having a serious advantage with respect to the classic model. In the classic model, the counter-current imbibition is described in terms of a parabolic equation which requires two boundary-values conditions at the inlet and at the outlet of the medium. At the same time, the inlet saturation cannot be imposed in the true process, as its value is established automatically. In the suggested model, the first-order differential equation does not require the inlet condition; the inlet saturation is established automatically according to the structure of the capillary flux function only.

6.4. DISPLACEMENT BY A WETTING FLUID

Examine the displacement process in terms of the problem (34) in the case when the capillary vector is directed along the positive sense of the axis x: $e_c = +1$. This means that the displacing phase is more wetting. The total flux $G(s)$ is then illustrated in Figure 10.

For small Ca, the solution is qualitatively similar to that presented in Figure 9(b) for the case of a capillary imbibition, with the negative front slower than the positive front, and the inlet saturation closer to 1. Two fronts are entirely determined by two tangent lines, OA and CB, constructed in Figure 10(a).

For large capillary numbers (without capillary forces), the problem is similar to the classical Buckley–Levertt problem. In this case, the solution contains a single front advancing into the medium. The disappearing of the beak front is explained by the fact that the high injection flow rate imposed at the inlet section eliminates any counter-current flow.

For moderate Ca, a new situation appears when the total flux G has the form shown in Figure 11(a).

The solution which is presented in Figure 11b contains two fronts, but both are now "positive", however their velocities are highly different from one other. The back front is much slower. Both fronts verify the Hugoniot–Lax–Oleinik conditions.

In all the cases, the inlet saturation is established automatically, which is conform to the physics of the process.

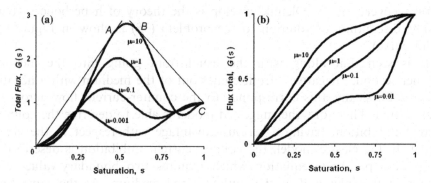

Figure 10. Total flux $G(s)$ in the case of wetting displacing phase: $Ca = 0.1$ (a) and $Ca = 1$ (b).

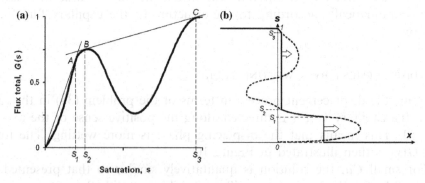

Figure 11. Graphical construction of the solution for moderate Ca: diagram $G(s)$ (a) and saturation profile (b).

6.5. DISPLACEMENT BY A NON-WETTING FLUID

In this case, the capillary vector is opposed to the flow direction: $e_c = -1$.

For large Ca, the capillary forces are neglected, and, thus, the wettability does not play any role. The saturation behaviour is close to that studied in the previous case of the wetting displacing phase.

For small Ca, another behaviour is observed. The total flux $G(s)$ is shown in Figure 12(a).

A critical capillary number, Ca_{cr}, arises, from which the function $G(s)$ has no more positive tangent lines starting from extremal points $(0, 0)$ and $(1, 1)$. This means that the problem has no discontinuous solutions describing the fluid penetration into the medium. This critical situation is shown in Figure 12(b). Physically, this means that below Ca_{cr} the displacing fluid cannot penetrate into the medium due to the capillary counter-force.

The critical capillary number Ca_{cr} is determined as the maximal value for which the following non-linear equation has solution s^- $(0 < s^- < 1)$:

$$G'(s^-) \equiv F'(s^-) - \frac{1}{Ca} F_c'(s^-) = \frac{F(s^-) - (1/Ca)F_c'(s^-)}{s^-},$$

which satisfies a supplementary condition: $G'(s^-) > 0$.

6.6. COMPARISON TO THE MICROSCALE NUMERICAL SIMULATIONS

The fluid displacement process with a wetting displacing phase was simulated using the network numerical algorithm described above. We studied a 3D flow in a parallelepiped. Initially, the medium was filled up with a single fluid. The displacing phase entered via a side and left the medium via the opposite side. All the other sides were impenetrable. Thus, the

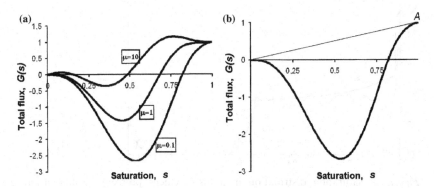

Figure 12. Total flux $G(s)$ for a non-wetting displacing phase and small Ca.

macroscale flow is one-dimensional and may be compared to the analytical 1D solutions obtained in Section 5.

The capillary number was 2, the viscosity ratio 0.5. The calculated saturation averaged over the cross-section is shown in Figure 13 (the solid line).

To construct the corresponding analytical solution, it is sufficient to determine the equivalent total flux $G(s)$. This function was calculated according to its definition: as the flow rate of the displacing phase divided to the total flow rate in various flow cross-sections characterized by various saturations s. In the examined case, we have obtained function $G(s)$ of the same qualitative type as in Figure 11(b).

The analytical solution which corresponds to this function $G(s)$ is shown in Figure 13.

A difference between two solutions which is visible in the vicinity of the advancing front is not a defect of the new model. It is simply the corollary of the fact that the analytical solutions are obtained for a homogeneous medium which does not cause the dispersion effect.

It is clearly seen that the numerical simulation manifests the effect of two fronts described in Section 6.4. This effect can not be captured within the framework of classic models.

A clearly observed defect of the analytical solution is the monotonous behaviour. It is not entirely consistent to the numerical curve, which has a minimum. Such a behaviour of the numerical curve reflects a more complicated interior process within the M-continuum than that determined by closure relations (21) and (20). In reality, the internal saturation s^I varies in time, which is the result of a mass exchange between various continua. This opens a perspective to improve the model.

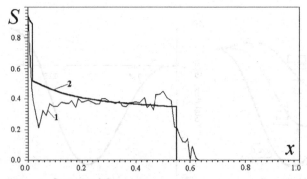

Figure 13. Saturation distribution in a displacement problem: results of numerical network simulations (oscillating curve) and analytical solution (monotonous curve).

6.7. QUALITATIVE RESULTS. COMPARISON TO THE CLASSIC MODEL

6.7.1. *Capillary Imbibition in a Homogeneous Medium*
Let us examine the capillary imbibition in a porous medium presented by a bundle of equivalent and uniform capillary tubes. The propagation of each meniscus in each tube is absolutely identical, so the macroscopic displacement front will have the form of a Heaviside step-function and will propagate as a travelling wave without any dissipation. Mathematically, this may only be obtained as the solution of a hyperbolic first-order equation. This is absolutely conform to the suggested model (34).

In contrast, this expected saturation profile can not be obtained within the framework of the classic Buckley–Leverett model which describes the capillary imbibition in terms of a parabolic diffusion equation.

6.7.2. *About a Square-Root Time Evolution of the Displacement Front*
The multiple experimental data obtained for a capillary imbibition (water penetrates into a medium saturated by air) show that the coordinate of the displacement front varies as a square-root of time. This is frequently used as a justification of the parabolic character of the modelling transport equation. For a linear hyperbolic process, the propagation of a front x_f is governed by the linear law: $x_f \sim t$.

At the same time, the square root dependence is not necessarily the property of a parabolic equation. This may be the result of a hyperbolic but non-linear character of the system. To illustrate such a possibility, it is sufficient to examine the meniscus propagation in a porous channel described by the Washburn equation (2). If the viscosity of the displaced phase is zero (the air) then the solution is

$$x_f \sim \sqrt{\frac{2P_c}{\mu_w}}$$

with a square-root dependence. At the same time, the system is purely hyperbolic, with no dissipation mechanisms.

Thus, the hyperbolic character of the capillary flow prescribed by the suggested model does not contradict the observed experimental data.

6.7.3. *About a Finite Velocity of Perturbation Propagation*
A supplementary advantage of the suggested model consists in the finite velocity of a front propagation. This is absolutely conform to the experimental observations. At the same time, the parabolic classic model has an infinite propagation velocity for any perturbations in general cases. Only in some particular situations, defined by a special type of nonlinearity, may a

parabolic equation have weakly discontinuous solutions with a finite propagation velocity (Barenblatt et al., 1990; Samarsky et al., 1995).

Thus the finite front velocity is a supplementary justification of the
meniscus model.

7. Conclusions

In the present paper, we have developed a model describing the fluid displacement in porous media, the process which is highly non-equilibrium
with respect to the local phase distribution. Two new constructive elements
of the two-phase flow field are suggested: the concept of a M-continuum
playing the role of a third phase, and the vectorial capillarity.

The M-continuum is presented by two fluids located in the vicinity of
menisci, examined as a unite phase possessing some particular dynamic
properties: a unite flow velocity common for both true phases and a
viscosity being a weighted mean value between two individual viscosities.

The vectorial capillarity concept results from the examination of the
meniscus flow inside a pore (5). For a meniscus structure, the capillary
forces play the role of true forces as a vectorial notion. They can accelerate
or decelerate the flow, by changing the system momentum. It is then natural that the capillarity appears in the momentum balance equation for the
M-continuum.

These two elements are sufficient to develop phenomenological two-
phase flow equations. To close the system we have suggested two supplementary relations which result from the microscale analysis of the meniscus
structure and numerical simulations based on a capillary network model of
porous media.

The mathematical form of the new model is qualitatively different from
the classic Buckley–Leverett model with capillary pressure. The suggested
model is hyperbolic of the first order in all the cases, even when the capillarity dominates. This is not traditional, as the classic model with capillary
forces is parabolic and, thus, of the second order. In the last two sections
of this paper we have shown, through the use of analytical solutions to test
problems and various physical examples that the hyperbolicity corresponds
well to the physical nature of the capillary meniscus movement.

Acknowledgements

This work is achived within the framework of the Groupement de Recher-
che MoMas: Modélisation Mathématique et Simulations Numériques Liées
aux Études d'Entreposage Souterrain de Déchets Radioactifs, CNRS,
ANDRA, BRGM, CEA, EDF.

References

Adler, P. M.: 1992, *Porous Media: Geometry and Transports*. Butterworth-Heinemann, New York, U.S.A.

Auriault, J. L. and Sanchez-Palencia, E.: 1986, Remarques sur la loi de Darcy pour les écoulements biphasiques en milieu poreux. *J. de Mec. Theor. Appl.* Special Issue, 141–156.

Barci, J. C., Chaouche, M. and Salin, D.: 1990, Modle simple de perméabilités relatives croisées. *C. R. Acad. Sci. Paris* **311** (série II) 591–597.

Barenblatt, G. I.: 1971, Flow of tow immiscible fluids in homogeneous porous media. *Izvestiia Akademii Nauk SSSR, Mekhanika Zhidkosti i Gaza* (5), 144–151.

Barenblatt, G. I. and Entov, V. M.: 1972, Non-equilibrium effects in the flow of immiscible fluids. *Numerical methods of solution to problem of multiphase incompressible fluid filtration, Periodical. Proc. Akademiia Nauk SSSR*, Novosobirsk, 33–43 (in Russian).

Barenblatt, G. I., Entov, V. M. and Ryzhik, V. M.: 1990, *Theory of Fluid Flows Through Natural Rocks*. Kluwer Academic Publishers, Dordrecht.

Barenblatt, G. I. and Vinichenko, A. P.: 1980, Non-equilibrium flow of immiscible fluids in porous media. *Uspekhi Mat. Nauk* **3**, 35–50.

Blunt, M. J. and King, P.: 1991, Relative permeabilities from two and tree dimensional pore scale network modelling. *Trans. Porous Media* **6**, 107–119.

Bocharov, O.: 1991, Influence of the non-equilibrium on two-phase flow in porous media. *l'coulement diphasique Akademiia Nauk SSSR, Proc. Ins. Teplofiziki*, Novosibirsk, 89–95.

Bourgeat, A. and Panfilov, M.: 1998, Effective two-phase flow through highly heterogeneous porous media: capillary nonequilibrium effects. *Comput. Geosci.* **2**, 191–215.

Buevich, U. A. and Mambetov, U. M.: 1991, Theory of flow simultaneous of immiscible fluids. *Inzhinerno-Fizicheskii Zhurnal* **60**, 98–107.

Entov, V. and Chen-Sin, E.: 1987, Micromechanics of two phase flow in porous media. *Numerical methods of problem solution of multiphase incompressible fluid filtration, Periodical. Proc. Akademii Nauk SSSR*, Novosobirsk, pp. 120–129 (in Russian).

Hassanizadeh, S. M.: 1997, Dynamic effects in the capillary pressure-saturation relationship, in: *Proceedings of the fourth International Conference on Civil Engineering*, Sharif Univ. Technol., Tehran, Iran, May 4–6, 1997, **4**, *Water Res. Environ. Eng.* 141–149.

Hassanizadeh, S. M. and Gray, W. G.: 1990, Mechanics and thermodynamics of multiphase flow in porous media including interface boundaries. *Adv. Water Res.* **13**, 169–186.

Hassanizadeh, S. M. and Gray, W. G.: 1993a, Toward an improved description of the physics of two-phase flow. *Adv. Water Res.*, **16**, 53–67.

Hassanizadeh, S. M. and Gray, W. G.: 1993b, Thermodynamic basis of capillary pressure in porous media. *Water Resour. Res.* **29**, 3389–3405.

Kalaydjian, F.: 1990, Origin and quantification of coupling between relative permeabilities for two-phase flow in porous media. *Trans. Porous Media*, **5**, 215–229.

Kalaydjian, F., Bourbiaux, B. and Cuerillot, D.: 1989, Viscous coupling between fluid phase for two-phase flow in porous media: theory versus experiment. in: *Proceedings of the fifth European Sympony on improved oil recovery*, Budapest, pp. 717–726.

Kurbanov A. K.: 1968, Equations of two-phase liquid transport in porous media. *Theory and Practice of Oil Reservoir Exploitation*, Nedra, Moscow, 281–286. (in Russian).

Lax, P. D.: 1971, Hyperbolic systems of conservation laws, II. *Communications of Pure Applied Mathematics*, **10**, Academic Press, New York, pp. 603–634.

Marinbah, M. A. and Lusternik, V. E.: 1985, Influence of the macroscopic heterogeneity on relative permeability of porous media. *Dynamics of Multiphase Media, Periodically Proceedings of the Academic Science*, Nauka, Novosobirsk, pp. 123–127 (in Russian).

Marle, C. M.: 1982, On macroscopic equations governing multiphase flow with diffusion and reactions in porous media *Int. J. Eng. Sci.* **20**, 643–662.

Nikolaevskii, V. N., Bondarev, E. A., Mirkin, M. I., Stepanova, G. and Terzi, V.: 1968, *Transport of hydrocarbure mixtures in porous media.* Nedra, Moscow (in Russian).

Oleinik, O. A.: 1959, Uniqueness and stability of the generalized solution of the Cauchy problem for a quasi-linar equation. *Uspekhi Mat. Nauk (N. S.)*, **14**, 165–170.

Panfilov, M.: 2000, *Macroscale Models of Flow Through Highly Heterogeneous Porous Media.* Kluwer Academic Publishing, Dordrecht.

Panfilova, I.: 2003, Modèle de ménisque d'écoulement diphasique en milieux poreux, PhD Thesis, Institut National Polytechnique de Lorraine, Nancy.

Panfilova, I., Shakirov, A., Kukharonok, A. and Panfilov, M.: 1998, Dynamic capillary model for two-phase flow through media with heterogeneous wettability. in: *Proceedings of the ECMOR-VI, Sixth European Conference on the Mathematics of Oil Recovery*, Peebles, near Edinbourg, Scotland, 8–11 September 1998, (paper C-09).

Panfilova, I. and Muller, J.: 1996, Two-scale method for the saturation transport simulation in the network model of porous media. in: *Proceedings of the International Congress "Recent Advances in Problems of Flow and Transport in Porous Media"*, Maroc, Marrakech, pp. 9–12, June 1996.

Panfilova, I. and Panfilov, M.: 2000, New model of two-phase flow through porous media in a vector field of capillary forces. in: *Porous Media: Physics, Models, Simulation, Proceedings of the International Conference*, World Scientific Publishing, Singapore, pp. 145–165.

Quintard, M. and Whitaker, S.: 1998, Two phase flow in heterogeneous porous media: the method of large-scale averaging. *Trans. Porous Media* **3**, 357–413.

Raats, P. A. C. and Klute, A.: 1968, Transport in soils: the balance of momentum. *Soil Sci. Sor. Amer. Proc.* **32**, 452–456.

Rhee, H., Aris, R. and Amundson, N. R.: 1986, *First-order Partial Differential Equations, Volume I*, Prentice-Hall, Englewood Cliffs, NJ.

Rose, A.: 1967, Model of doublet for simulation of trapping in porous media. *Trans. AIME* **112**(23), 345–356.

Rose, W.: 1989, Data interpretation problems to be expected in the study of coupled tow-phase flow of immiscible fluid flows in porous media. *Trans. Porous Media* **4**, 185–198.

Rothman, D. H.: 1989, Lattice-gas automata for immiscible two-phase flow. *Discrete kinetic theory, lattice-gas dynamics, and foundations of hydrodynamics.* Singapore, pp. 286–299.

Smaraski, A. A., Galaktionov, V. A., Kurdiumov, V. P. and Mikhailov, A. P.: 1995, *Blow-up in Quasilinear Parabolic Equations*, Vol. 19, de Gruiter Expositions in Mathematics, de Gruiter, Berlin and Hawthorne, NY.

Shvidler, M. I.: 1961, Two-phase flow equations in porous media providing for the phase interaction. *Izvestiia Akademii Nauk SSSR, Mekhanika, Mashinostroenie* (1), 31–134.

Singhal, A. K. and Somerton, W. H.: 1997, Quantitative modelling of immiscible displacement in porous media: network approach. *Rev. Inst. Fr. Petrol* **32**(6), 897–920.

Thigpen, L. and Berryman, J.: 1985, Mechanics of porous elastic materials containing multiphase fluid. *Int. J. Eng. Sci.* **23**, 1203–1214.

Vinichenko, A. P.: 1978, Theory of non stationary non-equilibrium flow of immiscible fluids. *Izvestiia Akademii Nauk SSSR, Mekhanika Zhidkosti i Gaza* (9), 51–58.

Whitaker, S.: 1986, Flow in porous media. 2: the governing equations for immiscible, two-phase flow. *Transport in Porous Media* **1**, 105–125.

Zarcone, C. and Lenormand, R.: 1994, Détérmination expérimentale du couplage visquux dans les écoulements diphasiques en milieu poreux. *C. R. Acad. Sci. Paris* **318**, (série II), 1429–1435.
Zarubin, U. A.: 1993, Exchange model of displacement in porous media. *Izvestiia Akademii Nauk SSSR, Mekhanika Zhidkosti i Gaza*, (1), 91–97.

Transp Porous Med (2005) 58:121–145
DOI 10.1007/s11242-004-5472-6

Macro-Scale Dynamic Effects in Homogeneous and Heterogeneous Porous Media

SABINE MANTHEY[1,*], S. MAJID HASSANIZADEH[2] and
RAINER HELMIG[1]
[1]*Institute of Hydraulic Engineering, Chair of Hydromechanics and Modeling of
Hydrosystems, Universität Stuttgart, Pfaffenwaldring 61, 70569 Stuttgart, Germany*
[2]*Department of Earth Sciences, Utrecht University, P.O. Box 80021, 3508 TA Utrecht,
The Netherlands*

(Received: 29 April 2003; in final form: 6 May 2004)

Abstract. It is known that the classical capillary pressure-saturation relationship may be deficient under non-equilibrium conditions when large saturation changes may occur. An extended relationship has been proposed in the literature which correlates the rate of change of saturation to the difference between the phase pressures and the equilibrium capillary pressure. This linear relationship contains a damping coefficient, τ, that may be a function of saturation. The extended relationship is examined at the macro-scale through simulations using the two-phase simulator MUFTE-UG. In these simulations, it is assumed that the traditional equilibrium relationship between the water saturation and the difference in fluid pressures holds locally. Steady-state and dynamic "numerical experiments" are performed where a non-wetting phase displaces a wetting phase in homogeneous and heterogeneous domains with varying boundary conditions, domain size, and soil parameters. From these simulations the damping coefficient τ can be identified as a (non-linear) function of the water saturation. It is shown that the value of τ increases with an increased domain size and/or with decreased intrinsic permeability. Also, the value of τ for a domain with a spatially correlated random distribution of intrinsic permeability is compared to a homogeneous domain with equivalent permeability; they are shown to be almost equal.

Key words: two-phase flow, dynamic capillary pressure-saturation relationship, macro-scale, damping coefficient

1. Introduction

For the simulation of two-phase flow processes, as e.g. the flow of an organic contaminant through an aquifer, a correctly determined constitutive relationship between saturation and capillary pressure plays a paramount role. Whereas fluid and structure properties such as porosity or permeability are well examined and known the determination of the

---

*Author for correspondence: e-mail: sabine.manthey@fak2.uni-stuttgart.de

fluid-structure interaction still poses a lot of questions. Thus during recent decades many experimental and methodical works have dealt with measuring or theoretically deriving this functional relationship. Beside hysteresis phenomena, researchers have identified a non-uniqueness arising from dynamic flow conditions. A review of experimental evidence from the literature is given by Hassanizadeh *et al.* (2002).

The influence of the flow conditions on capillary pressure (P_c)-saturation (S_w) relationships is illustrated with experiments carried out by Topp *et al.* (1967). They measured three P_c–S_w curves in the laboratory under a variety of flow conditions for quasi-static, steady-state and dynamic conditions (Figure 1). The curves for the quasi-static and the steady-state cases, where the rate of change of water saturation $\partial S_w/\partial t$ is zero at the shown data points, differ only slightly. In contrast, the dynamic curve, arrived at by increasing the gas pressure at the upper end of an initially fully water-saturated soil sample within $\Delta t = 100\,\mathrm{min}$ from $P_n = 0\,\mathrm{Pa}$ to $P_n = 550\,\mathrm{Pa}$, lies above the two other ones. Consequently, the relationship between saturation and capillary pressure cannot be regarded as unique. Many other experimental results have shown a dependence on the flow conditions, too (see Hassanizadeh *et al.*, 2002 for reference). With an increasing pressure gradient, induced at the boundaries, the difference between the equilibrium and the dynamic P_c – S_w relationship becomes larger.

In numerous applications, such as the decontamination of a NAPL spill, in oil recovery, or technical applications such as paper production or filter technology large gradients may occur. For the simulations of the two-phase flow processes, nevertheless, unique relationships between capillary pressure

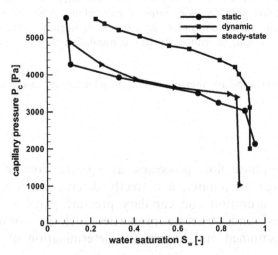

Figure 1. Three P_c–S_w relationships (primary drainage) measured in the laboratory under static, steady-state and dynamic conditions after Topp *et al.* (1967).

and water saturation are assumed most of the time. To deal with the non-uniqueness described above and to be able to incorporate the dynamic effects into numerical models, some functional relationship has to be provided. Furthermore, parameters for this relationship need to be determined.

In this paper, first we will introduce different model concepts dealing with non-equilibrium effects in two-phase flow, choosing one of them to examine further. The study is carried out by means of numerical simulations that resemble laboratory experiments for the measurement of capillary pressure-saturation relationships. At the local scale, it is assumed that Darcy's law and the traditional capillary pressure theory hold.

2. Extended Capillary Pressure-Saturation Relationships

In the following, different model concepts will be described which capture non-equilibrium effects in the constitutive relationships of two-phase flow.

Hassanizadeh and Gray (1993) volume averaged pore scale balance equations for mass, momentum, energy and entropy. The authors arrived at a functional relationship between the rate of change of water saturation, $\partial S_w / \partial t$, and the difference between the pressures of the non-wetting phase, P_n, the wetting phase, P_w, and the capillary pressure at equilibrium, P_c^{eq}, (static or steady-state) at a given saturation. As a first approximation, they linearised the relationship, assuming that under static or steady-state conditions the rate of change of saturation and thus the difference between the three pressures would be zero (see Equation (2.1)). The negative sign on the right hand side ensures that the entropy production is always larger than zero.

$$P_n - P_w - P_c^{eq}(S_w) = -\tau \frac{\partial S_w}{\partial t}. \tag{2.1}$$

Here τ can be regarded as a damping or relaxation coefficient, in this paper we will call it a damping coefficient. For the sake of brevity, the rate of change of water saturation $\partial S_w / \partial t$ will be abbreviated to saturation rate.

Under dynamic flow conditions the difference in the phase pressures may not equal the equilibrium capillary pressure. For the sake of brevity, we shall refer to $P_n - P_w$ as "dynamic capillary pressure", such that Equation (2.1) becomes:

$$P_n - P_w = P_c^{dyn} = -\tau \frac{\partial S_w}{\partial t} + P_c^{eq}(S_w). \tag{2.2}$$

This equation suggests that a new equilibrium between the dynamic and the equilibrium capillary pressure is not attained instantaneously. The damping coefficient might depend on the water saturation. The Equation (2.2) will be referred to as the extended P_c-S_w relationship in the following.

Based on experimental evidence Stauffer (1978) came up with a similar relationship suggesting that the damping coefficient τ can be quantified for different soils as given in Equation (2.3).

$$\tau = \frac{\alpha \cdot \phi \cdot \mu_w}{\lambda \cdot K} \left(\frac{P_d}{\rho_w \cdot g} \right)^2. \tag{2.3}$$

In Equation (2.3) α is a dimensionless parameter (assumed to be $\alpha = 0.1$), ϕ is the porosity and K is the intrinsic permeability; μ_w is the dynamic viscosity and ρ_w the density of water; P_d and λ are the Brooks and Corey parameters (see Section 3.1 and Equation (3.5)).

Stauffer (1978) also examined dynamic effects in the relative permeability-saturation relationship but found them to be of minor importance for the analysed cases.

Concluding from pore scale flow processes Barenblatt et al. (2002) state that during imbibition processes non-equilibrium effects arise due to redistribution processes on the pore scale. During imbibition the relative permeability of the wetting phase might then be higher than the one evaluated at the actual water saturation due to the system's relaxation. They, therefore, suggested the introduction of an apparent water saturation η, which differs from the actual saturation S_w as defined in Equation (2.4). The constitutive relationships, which are not altered, would be evaluated based on the value of η instead of the actual water saturation. For example, for imbibition processes the apparent saturation of the wetting phase would be larger than the actual water saturation, consequently, the relative permeability k_{rw} would be higher than if evaluated based on S_w. This is in agreement with laboratory observations. At equilibrium the saturation η equals the actual water saturation. In order to define a relation between the non-equilibrium and the actual saturation the authors assume a functional dependence on the relaxation time τ_B and the saturation rate arriving at Equationn (2.4) after a dimensional analysis.

$$\eta - S_w = \tau_B \frac{\partial S_w}{\partial t}. \tag{2.4}$$

The authors not only deal with non-equilibrium effects in the capillary pressure-saturation relationship but link it with a non-equilibrium relative permeability-saturation relationship.

Barenblatt et al. (2002) remark that non-equilibrium effects should be taken into account for fast flow processes whose time scale is small compared to the relaxation time τ_B. For slow flow processes, whose time scale is in the order of the relaxation time, the assumption of local equilibrium is justified. The authors have applied their model in numerical calculations, as did Silin and Patzek (2004) with their study on the Barenblatt model.

Bourgeat and Panfilov (1998) applied homogenisation theory to the balance equations for two immiscible fluids in a porous medium with periodical heterogeneities made up of a coarse background material (superscript c) with embedded fine lenses (superscript f). The water saturation rate within the fine material, damped by the coefficient τ_{BP}, is found to be a function of the difference in capillary pressures of the fine and coarse material (see Equation (2.5)).

$$\frac{1}{Ca}P_c^f\left(S_w^f\right) - P_c^c\left(S_w^c\right) = -\tau_{BP}\frac{\partial S_w^f}{\partial t^*} \quad \text{with} \quad Ca = \frac{\Delta p_w}{p_c}. \tag{2.5}$$

In contrast to the coefficient τ_B of Barenblatt et al. (2002), which has the dimension of time, the τ_{BP} parameter of Bourgeat and Panfilov has the same dimension ($MT^{-1}L^{-1}$) as τ in Equation (2.1).

Equation (2.5) is somewhat similar to Equation (2.2) suggested by Hassanizadeh and Gray (1993). In this analogy, $P_c^c\left(S_w^c\right)$, the capillary pressure of the coarse material may be considered to be similar to the equilibrium capillary pressure, as the equilibrium in the coarse material may be reached rather fast. The capillary pressure of the fine sand, $P_c^f\left(S_w^f\right)$, may considered to be similar to the dynamic capillary pressure as equilibrium will be attained slower.

The approach of Bourgeat and Panfilov (1998) is limited to periodic heterogeneities made up of two kinds of materials. However, in the balance equations of two-phase flow we would prefer a more universal approach, that can be included in the balance equations for two phases easily.

For this paper we will more closely examine the relationship proposed by Hassanizadeh and Gray (1993) as it is applicable to all kinds of heterogeneity pattern and has not been analysed in detail on the macro-scale. Also, for the moment, we do not want to consider possible dynamic effects in the relative permeability-saturation relationship. It must be noted that the dynamic effect considered here is present only for transient flow, e.g. when the saturation rate is unequal to zero. Thus, cases where P_c may depend on the flow velocity (see e.g. DiCarlo and Blunt, 2000), which might also be the case under steady-state conditions, are not dealt with here.

Before applying the linear relationship between saturation rate and the difference in pressures as proposed by Hassanizadeh and Gray (1993) the linearisation has to be tested. Furthermore, functional dependencies, e.g. of τ on the water saturation, the soil parameters, or the domain size, have to be determined. We will study these issues in this paper based on numerical experiments with a continuum-scale model.

For work determining τ on the pore scale we refer to Dahle et al. (this book), Gielen et al. (2003, 2004), Hassanizadeh et al. (2002), and Singh and Mohanty (2003).

3. Numerical Experiments

In order to investigate the linear relationship (Equation (2.2)), numerical experiments are performed on a number of computational domains, with dimensions between 0.03 m and 1.0 m. The domains may be homogeneous or heterogeneous with respect to the permeability and/or entry pressure distribution.

At first, the equilibrium capillary pressure-saturation relationship for a given domain is determined. In a second step, dynamic numerical drainage experiments are performed where for each time step the domain averaged phase pressures and the saturation rate are computed. Using these data in Equation (2.2), we can then calculate the value of the damping coefficient τ as a function of water saturation.

It must be emphasized that it is not the purpose of this paper to show that a heterogeneous domain can be replaced with a homogeneous domain with effective parameters. In other words, we are not performing a full upscaling of equations. Instead, we are assuming that Equation (2.1) holds at all various scales and calculate τ at various scales. This is similar to the concept of dynamic viscosity. When upscaling turbulence, the same phenomenological equation is assumed to apply and the corresponding coefficient, dynamic viscosity, is assumed to grow with (time) scale.

In this section, the underlying physical-mathematical model, the numerical experiments (initial and boundary conditions, soil parameters), and the averaging procedures will be described.

3.1. PHYSICAL-MATHEMATICAL MODEL

The governing equations of two-phase flow in a porous medium on scales larger than the pore scale for an incompressible wetting phase (subscript w) and an incompressible non-wetting phase (subscript n) are given as below (see Equations (3.1) and (3.2)). Gravity and the contribution of source/sink terms are neglected. The solid phase is regarded as rigid and the intrinsic permeability K as isotropic.

$$\phi\rho_w\frac{\partial S_w}{\partial t} - \nabla \cdot \left\{ \rho_w \frac{k_{rw}(S_w)}{\mu_w} K \nabla P_w \right\} = 0, \tag{3.1}$$

$$\phi\rho_n\frac{\partial S_n}{\partial t} - \nabla \cdot \left\{ \rho_n \frac{k_{rn}(S_n)}{\mu_n} K \nabla P_n \right\} = 0 \tag{3.2}$$

In Equations (3.1) and (3.2) the primary variables are the saturations S_w and S_n and the pressures P_w and P_n. Φ denotes the porosity, ρ the density, μ the dynamic viscosity, k_r the relative permeability, and t the time. These two equations are strongly coupled by the relative permeabilities and the two restrictions given in Equations (3.3) and (3.4).

$$S_w + S_n = 1.0, \tag{3.3}$$

$$\nabla P_c = \nabla P_n - \nabla P_w \quad \text{with} \quad P_c = P_c(S_w). \tag{3.4}$$

The capillary pressure is assumed to be a unique function of the water saturation (Equation (3.4)). It will be parameterised with the Brooks and Corey (1964) relationship (see Equation (3.5)). Consequently, in our simulations local dynamic effects are neglected. The relative permeability-saturation relationships are calculated after the approach of Burdine (1953) based on the Brooks and Corey P_c–S_w relationship (Equation (3.6)).

$$P_c(S_e) = P_d S_e^{-1/\lambda}, \quad S_e = \frac{S_w - S_{wr}}{1.0 - S_{wr} - S_{nr}}, \tag{3.5}$$

$$k_{rw} = S_e^{2+3\lambda/\lambda}, \quad k_{rn} = (1.0 - S_e)^2[1.0 - S_e^{2+\lambda/\lambda}]. \tag{3.6}$$

Here S_e denotes the effective water saturation. Alternatively, we should have employed the extended capillary pressure-saturation relationship after Equation (2.1). The value of the damping coefficient would have been unknown. Typical values could have been chosen on the basis of pore-network model simulations. Values determined by Gielen et al. (2004) are less than $\tau = 10^5$ Pa s. Dahle et al. (2003) calculated τ-values in the order of $\tau = 10^2$ Pa s for a bundle of capillary tubes model. The effect on the solution of these values are expected to be small (Hassanizadeh et al., 2002). For now, we choose τ to be zero for the numerical calculations.

The balance equations are solved with the multiphase flow simulator MUFTE-UG (Bastian, 1999) based on a node-centered Finite-Volume discretisation in space, the so-called BOX-method (Helmig, 1997; Huber and Helmig, 1999), and a backward-difference Euler scheme for the time discretisation. A quasi-Newton–Rhapson algorithm solves the nonlinear system of equations. The linear system of equations evolving during the Newton–Rhapson algorithm is handled with a BiCGStab (stabilised biconjugate gradient) scheme applying a V-multigrid cycle as pre- and post-smoother.

3.2. INITIAL AND BOUNDARY CONDITIONS

The damping coefficient τ for a given sample may be determined if P_c^{dyn}, P_c^{eq} and the saturation rate $\partial S_w/\partial t$ are known at a given saturation. To obtain such data, two kinds of numerical experiments have to be carried out: steady-state (yielding P_c^{eq} for a given water saturation) and dynamic experiments (yielding P_c^{dyn} and $\Delta S_w/\Delta t$). The corresponding simulations were set up to mimic a flow-through cell.

In a flow-through cell experiment initially the sample is assumed to be fully saturated with water. The wetting phase pressure varies linearly (from P_w to 0) from top to bottom and the non-wetting phase pressure is set to

Table I. Boundary conditions for the numerical experiments mimicking the flow through cell

Boundary	$t = 0$ s both exp.	$t > 0$ s steady-state	$t > 0$ s dynamic
Top (Dirichlet)	$P_w = \{0 \text{Pa},$ $10^2 \text{Pa}, 10^3 \text{Pa}\}$ $P_n = P_w + P_d$	$P_w = \{0 \text{Pa}, 10^2 \text{Pa}, 10^3 \text{Pa}\}$ $P_n = P_w + P_c^B$ with $P_c^B = P_d + \sum_l 1.5^l$	$P_w = \{0 \text{Pa}, 10^2 \text{Pa}, 10^3 \text{Pa}\}$ $P_n = P_w + P_c^{B,\text{Max}}$ with $P_c^{B,\text{Max}} = \{2.0 \cdot 10^3 \text{Pa},$ $4.0 \times 10^3 \text{Pa}, 1.0 \times 10^4 \text{Pa}\}$
Bottom (Dirichlet)	$P_w = 0$ $P_n = P_d$	$P_w = 0 \text{Pa}$ $P_n = P_c^B$ with $P_c^B = P_d + \sum_l 1.5^l$	$P_w = 0 \text{Pa}$ $P_n = P_c^{B,\text{Max}}$ with $P_c^{B,\text{Max}} = \{2.0 \times 10^3 \text{Pa},$ $4.0 \times 10^3 \text{Pa}, 1.0 \times 10^4 \text{Pa}\}$
Sides (Neumann)	$q_w = q_n$ $= 0 \text{ kg/(m}^3 \cdot \text{s})$	$q_w = q_n = 0 \text{ kg/(m}^3 \cdot \text{s})$	$q_w = q_n = 0 \text{ kg/(m}^3 \cdot \text{s})$

equal the wetting phase pressure plus the entry pressure P_d. The boundary conditions are imposed on the sample (read the modelling domain) as given in Table I and depicted in Figure 2.

Here P_c^B denotes the boundary capillary pressure. It is evident that a pressure difference of ΔP within both the wetting and the non-wetting fluids across the sample is imposed. In the steady-state experiment, the boundary capillary pressure P_c^B is increased stepwise following a geometric series. ΔP is kept constant. This means that only the non-wetting phase pressure is increased. Simulations are run until equilibrium

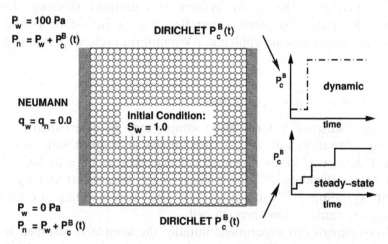

Figure 2. Boundary conditions for the steady-state and dynamic numerical experiments.

(i.e. steady-state flow for both fluids from top to bottom) is established for a given boundary capillary pressure. To determine whether equilibrium is reached the change of the averaged water saturation (see Equation (3.7)) and averaged water pressure (see Equation (3.9)) between two subsequent time steps is less than a residual of $\varepsilon_S < 1.0 \times 10^{-6}$ for the water saturation and $\varepsilon_P < 1.0 \times 10^{-2}$ Pa for the water pressure. Then the index 1 of the geometric series is by one (see Table I) and a new boundary capillary pressure is imposed. The average water saturation of the sample at equilibrium and the corresponding P_c^B provide one data set for the equilibrium P_c–S_w function.

In the dynamic experiment the boundary capillary pressure is raised at once to a high value $P_c^{B,Max}$. As a result, the non-wetting phase saturation in the sample rises quickly. After each time step average saturation and average phase pressures are calculated. These results are used to obtain $P_c^{dyn} = P_n - P_w$ and the saturation rate as a function of average saturation. Note that here, contrary to steady-state experiments, the boundary capillary pressure P_c^B is not representative of the average capillary pressure P_c^{dyn} at intermediate times (see Figure 5). In the dynamic experiment counter-current flow may occur.

3.3. MATERIALS AND PROPERTIES

In order to analyse the extended P_c–S_w relationship, numerical experiments were performed with four different set-ups:

- Homogeneous coarse sand (domain size $0.12 \, \mathrm{m} \times 0.12 \, \mathrm{m}$); this is effectively one-dimensional flow.

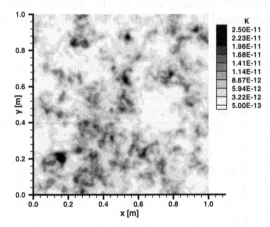

Figure 3. Simple heterogeneity pattern (left) and intrinsic permeability distribution for spatially correlated isotropic random field (right).

- Simple heterogeneity pattern (domain size 0.12 m × 0.12 m) made up of a background material (coarse sand) with two embedded fine sand blocks (see Figure 3, left).
- Heterogeneous permeability distribution based on an isotropic spatially-correlated random field (domain size 1.0 m × 1.0 m, see Figure 3, right) but with homogeneous entry pressure.
- Heterogeneous permeability and entry pressure distribution based on an isotropic spatially-correlated random field (domain size 1.0 m × 1.0 m).

The soil parameters of the coarse and fine sand are listed in Table II, where also the statistical parameters of the random field are given. For the parametrisation of the equilibrium P_c–S_w relationship, the Brooks and Corey (1964) (BC) formula was chosen. For the relative permeability-saturation relationship the Burdine (1953) formula was applied. The non-wetting phase (NAPL) viscosity ($\mu_n = 0.9 \times 10^{-3}\,\mathrm{kg\,m^{-1}\,s^{-1}}$) is 10% less than the wetting phase (water) viscosity ($\mu_n = 1.0 \times 10^{-3}\,\mathrm{kg\,m^{-1}\,s^{-1}}$).

The isotropic random heterogeneity field was generated based on an approach by Dykaar and Kitanidis (1992) with an exponential covariance function. As the grid consisted of 101^2 nodes, one correlation length ($a = 0.05\,\mathrm{m}$) was covered by five nodes, which means there are 20 correlation lengths in the x- and y-directions.

3.4. REPRESENTATIVE ELEMENTARY VOLUME (REV)

Before calculating the parameters for different heterogeneous set-ups, one needs to establish that a Representative Elementary Volume (REV) for the domain considered here exists. For the simple heterogeneous set-up (see Figure 3, right) this was shown by Ataie-Ashtiani et al. (2001). For the random heterogeneity field, the arithmetic means of the logarithm of the intrinsic permeabilities were calculated for four different starting points and

Table II. Parameters of the random field as well as the coarse and fine sands used for the simple heterogeneity pattern

Parameter	Random field	Coarse sand	Fine sand
Porosity ϕ [–]	0.4	0.4	0.4
Intrinsic Permeability K [m^2]	5.11×10^{-12}	5.0×10^{-9}	5.0×10^{-12}
BC P_d [Pa]	5.0×10^2	3.5×10^2	1.0×10^3
BC λ [–]	2.0	3.0	3.0
Res. Water Saturation S_{wr} [–]	0.05	0.08	0.1
Variance σ^2 of ln K	0.5		
Correlation length a [m]	0.05		

subsequently increasing averaging domains (see Figure 4). For the starting point 1, the final value does not yet equal the average of the ln K distribution. For all other starting points, an REV is reached even before the whole domain is taken into account. Subsequently the averaging of the pressures and the saturations is performed over the whole domain including boundary nodes for the simple heterogeneity pattern and the spatially correlated random field.

3.5. AVERAGE PRESSURES AND SATURATIONS

To arrive at averaged parameters for the extended P_c–S_w relationship average phase pressures and water saturations after each time step at time level t during the numerical experiments have to be computed. The average water saturation $\bar{S}_w(t)$ is the weighted arithmetic mean (Equation (3.7)).

$$\bar{S}_w(t) = \sum_{i=1}^{m} S_{w,i}(t) \cdot w_i \quad \text{with} \quad w_i = \frac{V_i}{V_{\text{dom}}} \tag{3.7}$$

V_i designates the volume belonging to a node of the node-centred Finite-Volume grid consisting of m nodes. V_{dom} denotes the volume of the whole domain. As we have a node-centered grid the volumes belonging to a node differ even for a regular grid. But basically, Equation (3.7) results in an arithmetic mean of the saturation. Based on the averaged water saturations, the saturation rate is calculated. It is approximated based on a central difference scheme (Equation (3.8)) for each time level t at the time step n.

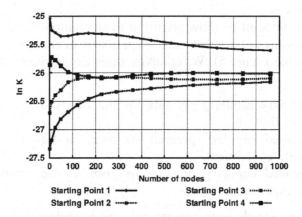

Figure 4. Arithmetic average of ln$K$ as a function of values included for the heterogeneous permeability distribution (starting point 1: $x=0.3$, $y=0.3$, starting point 2: $x=0.6$, $y=0.3$, starting point 3: $x=0.3$, $y=0.6$, starting point 4: $x=0.6$, $y=0.6$).

$$\frac{\Delta \bar{S}_w^{t^n}}{\Delta t} = \frac{\bar{S}_w^{t^{n+1}} - \bar{S}_w^{t^{n-1}}}{t^{t^{n+1}} - t^{t^{n-1}}}. \tag{3.8}$$

In many experimental set-ups or when applying the method of capillary equilibrium the boundary pressures of the phases yield an average capillary pressure. Whereas this is an appropriate choice for static or steady-state conditions it is not applicable to dynamic experiments. Here, the average phase pressures of the wetting phase $\langle P_w \rangle$ and the non-wetting phase $\langle P_n \rangle$ at time t are calculated after Equatin (3.9), where the phase pressure at each node i is weighted by the volume $V_{\alpha i}$ of the phase at node i.

$$\langle P_\alpha \rangle(t) = \frac{\sum_i^m P_{\alpha i}(t) \cdot V_{\alpha i}(t)}{\sum_i^m V_{\alpha i}(t)}, \quad V_{\alpha i}(t) = S_{\alpha i}(t) \cdot \phi_i \cdot V_i \quad \text{with} \quad \alpha = w, n. \tag{3.9}$$

Based on the average phase pressures the average capillary pressure can be calculated after Equation (3.10).

$$\langle P_c \rangle(t) = \langle P_n \rangle(t) - \langle P_w \rangle(t). \tag{3.10}$$

4. Interpretation of the Simulation Results

In the following we will show the variations of the average capillary pressures and the average water saturation rate as a function of the water saturation. These two quantities provide the basis for a regression minimizing the sum of squared differences between the extended P_c–S_w relationship and data points.

But, first we want to draw attention to the temporal development of the capillary pressures of a steady-state experiment in comparison to a dynamic experiment (see Figure 5). We compare three capillary pressures as function of time:

- The boundary capillary pressure for the steady-state experiment P_c^B as defined in Table I.
- The averaged capillary pressure as given in Equation (3.10) for the steady-state experiment.
- The averaged capillary pressure as given in Equation (3.10) for the dynamic experiment.

During the steady-state experiment, the boundary capillary pressure is a good approximation (at least for the first eight pressure steps) of the difference in the averaged phase pressures $\langle \bar{P}_n \rangle - \langle \bar{P}_w \rangle$. On the contrary, the

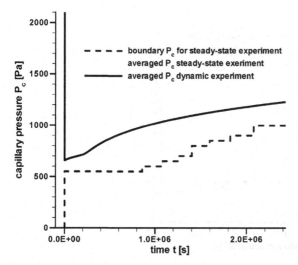

Figure 5. Capillary pressures as functions of time for a steady-state and a dynamic ($P_c^B(t > 0) = 4.0 \times 10^3$ Pa) experiment with the spatially-correlateed random heterogeneity field (see Figure 3 and Table II).

averaged capillary pressure does not equal the boundary capillary pressure during the complete dynamic experiment.

A plot of the difference between the average phase pressures as a function of saturation for the simple heterogeneity pattern is shown in Figure 6. The very large dynamic capillary pressure in the vicinity of $S_w = 1.0$ (corresponding to the initial time steps of the computations) is a numerical artefact. If the first time step were increased from $\Delta t = 0.001$ s to $\Delta t > 0.005$ s, this local maximum in the dynamic capillary pressure would not be noticed. The question remains how to identify the extended P_c–S_w relationship in a consistent manner at water saturations of $S_w \to 1.0$. In the following analysis, the first time steps are neglected when examining the dynamic numerical experiments.

Next the difference between the dynamic and equilibrium capillary pressures $\Delta P_c = \langle P_c^{dyn} \rangle - \langle P_c^{equil} \rangle$ and the rate of change of saturation are plotted as functions of water saturation (see Figure 7). The difference between the dynamic and equilibrium capillary pressures as a function of the water saturation is not monotonously decreasing. It has a local maximum at a water saturation $\bar{S}_w \cong 0.78$. For saturations less than this, the difference decreases monotonously for decreasing water saturations. At the same time, the saturation rate is monotonously decreasing for all water saturations. The non-monotonous behaviour of the difference between dynamic and equilibrium capillary pressures is not caused by the large difference in the phase pressures during the first two time steps. These are not taken into account. A similar behaviour can be observed for different heterogeneous set-ups. This

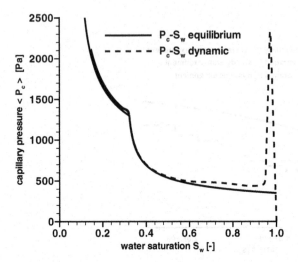

Figure 6. Difference of the averaged phase pressures of the steady-state and the dynamic ($P_c^B = 4.0 \times 10^3$ Pa, $\Delta P = 100$ Pa) experiment for the simple heterogeneity pattern.

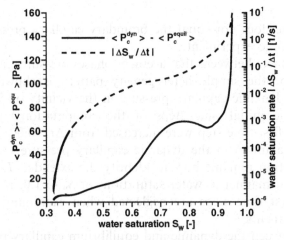

Figure 7. Saturation rate and difference between the dynamic and equilibrium capillary pressure as a function of water saturation for a dynamic numerical experiment with the simple heterogeneity pattern ($P_c^B = 4.0 \cdot 10^3$ Pa, $\Delta P_{w,n} = 100$ Pa).

is probably related to the fact that at an averaged water saturation of $\bar{S}_w \cong 0.7$ the fine sand lenses are still fully water saturated.

From these curves a plot of the difference between the dynamic and equilibrium capillary pressure vs. saturation rate at a fixed saturation can be prepared to determine τ and to test the validity of Equation (2.1). A typical data set is shown in Figure 8 for three different dynamic experiments with boundary capillary pressures of $P_c^B = 2.0 \times 10^3$ Pa,

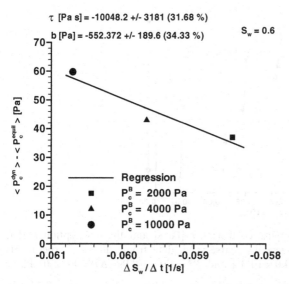

Figure 8. Regression based on the linear relationship (Equation 4.1) for three different dynamic experiments at a water saturation of $S_w = 0.6$ for the homogeneous
coarse sand ($\Delta P = 1.0 \times 10^2$ Pa).

$P_c^B = 4.0 \times 10^3$ Pa, $P_c^B = 10^4$ Pa for a water saturation of $\bar{S}_w = 0.6$. Such a
plot, according to Equation (2.1) should indicate a linear relationship pass-
ing through the origin. The slope of this line equals the damping coefficient
τ.

A regression minimising the sum of the squared differences between the
data and the extended P_c–S_w function was performed (least-squares Leven-
berg-Marquard algorithm) based on the formula

$$\Delta P_c = P_c^{\mathrm{dyn}} - P_c^{\mathrm{eq}}(S_w) = b - \tau \frac{\partial S_w}{\partial t}, \tag{4.1}$$

where b represents the y-axis intercept.

In this way, τ can be determined for various water saturations. This has
resulted in the graph plotted in Figure 9 where $\Delta P_c(\Delta S_w / \Delta t)$ is shown
for $\bar{S}_w = 0.3, 0.4, \ldots, 0.8$. From Figure 9 it is again obvious, that the linear
functions do not run through the origin, there is an intercept b.

The values determined for τ and b depend on the water saturation (see
Table III). The value of the damping coefficient, τ, increases with decreas-
ing water saturation from $\tau = 1.4 \times 10^3$ Pa s at $\bar{S}_w = 0.8$ to $\tau = 9.4 \times 10^5$ Pa s
at $\bar{S}_w = 0.2$. A non-linear relationship can be discerned. The values of b also
show this trend although not as clearly as in the case of τ.

Clearly three data points for one regression are not enough to test a
linear relationship. However, especially the highly negative values for b
suggest, that the linear relationship is at least questionable, because the

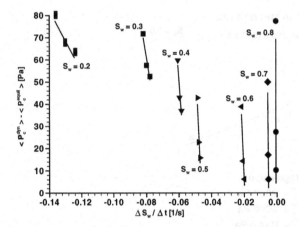

Figure 9. Values for damping coefficient τ [Pa s] and the interception b [Pa] with the y-axis for the homogeneous coarse sand, data from numerical experiments with $P_c^B = 2.0 \times 10^3$ Pa, $P_c^B = 4.0 \times 10^3$ Pa and $P_c^B = 1.0 \times 10^4$ Pa, $\Delta P = 1.0 \times 10^2$ Pa.

Table III. Values of τ and b for given water saturations for homogeneous coarse sand, data from numerical experiments with $P_c^B = 2.0 \times 10^3$ Pa, $P_c^B = 4.0 \times 10^3$ Pa and $P_c^B = 1.0 \times 10^4$ Pa, $\Delta P = 1.0 \times 10^2$ Pa

$\bar{S}_w$	τ [Pa s]	b [Pa]
0.2	9.4×10^5	-606
0.3	9.2×10^4	-478
0.4	1.8×10^4	-360
0.5	1.6×10^4	-774
0.6	1.0×10^4	-552
0.7	4.5×10^3	-303
0.8	1.4×10^3	-109

difference in the pressures should always be positive for drainage processes. A value of b smaller than zero means that at a saturation rate of zero the difference in phase pressures and equilibrium capillary pressure equals -552 Pa when it is expected to be zero. Also, at equilibrium between the phase pressures and the equilibrium capillary pressure the saturation rate would be (in this case) negative. Dahle *et al.* (this book) also determine b-values unequal to zero based on a bundle of capillary tubes model.

Kalaydjian (1992) calculated values of $\tau = 2 \times 10^6$ Pa s for a water–oil system under imbibition conditions using a constant influx in a sandstone with an intrinsic permeability of $K = 2.1 \times 10^{-13}$ m^2 and an entry pressure of approximately $P_d = 2.5 \times 10^3$ Pa. In our case the intrinsic permeability is a factor of 10^4 larger and the entry pressure is a factor of 10 smaller.

According to Stauffer's formula (Equation (2.3)) the value of τ should be decreasing for increasing permeability and decreasing entry pressure. These trends are confirmed by the results of the simulations compared to the results of obtained Kalaydjian (1992).

In the following the damping coefficient τ will be determined based on the assumption that the linear relationship holds for all ranges of the saturation rate and that the extended P_c–S_w relationship runs through the origin. This assumption reduces the number of simulations required significantly and should show trends in the $\tau\,(S_w)$ even if the relationship cannot be regarded as linear for all saturation rates.

5. Influence of boundary conditions

In order to show the influence of the boundary conditions we compare here the results of three simulations for the simple heterogeneous set-up applying three different boundary capillary pressures (see Figure 10). For water saturations $\overline{S_w} > 0.7$ the difference in the functions $\tau\left(\overline{S_w}\right)$ is not pronounced, implying that a linear relationship could hold for high water saturations. However, for smaller water saturations the functions differ. This can be interpreted as a functional dependence of τ on the saturation rate.

In order to interpret the results, the influence of the driving forces can be contemplated, e.g. the capillary number determines dominating forces of a two-phase flow system. The application of the capillary number would require a steady-state flow field. We will here assume that for a given water

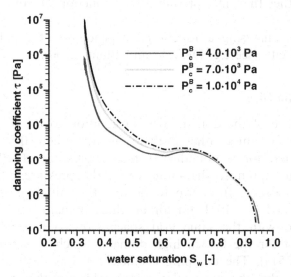

Figure 10. Damping coefficient τ (on log scale) as a function of the water saturation S_w for three boundary capillary pressures ($\Delta P = 100\,\mathrm{Pa}$).

saturation we can compare the saturation rate (and thus the flow velocities) of the three different experiments although steady-state is not given. The capillary number can be defined according to Equation (5.1) after Hilfer (1996), where v_w denotes the flow velocity of the wetting phase, L denotes a characteristic length and P_c^* denotes a typical capillary pressure of the system. In the three numerical experiments described above, the viscosity, the characteristic length and intrinsic permeability K do not differ.

$$Ca = \frac{\text{viscous forces}}{\text{capillary forces}} = \frac{\mu_w \cdot v_w \cdot L}{P_c^* \cdot K}. \tag{5.1}$$

For the influence of the capillary forces, the typical capillary pressure, we can apply the boundary capillary pressure. We may assume here the saturation rate to be proportional to the flow velocity of the wetting phase. In the numerical experiments the saturation rate increases with increasing boundary capillary pressure. Thus, the viscous as well as the capillary forces would increase which complicates the interpretation of the results. One should apply Neumann boundary conditions to make the conditions less ambiguous. Kalaydjian (1992) determined values for $\tau(S_w)$ for two imbibition processes varying the influx. For higher flow velocities he obtained smaller τ values as compared to τ values of a laboratory experiment inducing lower flow velocities. Assuming that the capillary forces can be considered as equal for the two experiments, τ would decrease with increasing viscous forces.

Summarising, we would like to state, that τ should rather be interpreted as a function of water saturation, of soil and fluid parameters, and perhaps of the saturation rate rather than interpreting it as a function of dominating forces.

There are approaches analysing a functional dependence of capillary pressure on the flow velocity (see e.g. DiCarlo and Blunt, 2000).

6. Influence of the Domain Size

To investigate the influence of the domain size, in addition to the 0.12 m high domain studied so far, three other domains namely (0.03, 0.5 and 1.0 m high) are considered for a boundary capillary pressure of $P_c^B = 4.0 \times 10^3$ Pa. In order to minimise the influence of the pressure difference within the phase pressures from top to bottom the initial pressure gradient was set to $\nabla P = 8.33 \times 10^3$ Pa/m for all three domain sizes. All domains were discretised with the same grid spacing of $\Delta y = 0.0025$ m in the y-direction, except for the 0.03 m high domain, which was discretised finer ($\Delta y = 0.000625$ m). The damping coefficient τ was determined based on the assumptions that the linear relationship holds and that it runs through the origin (see Equation (2.1)).

With increasing domain size, the value of the damping coefficient increases (see Figure 11) This is to be expected, as for increasing domain size the saturation rate decreases and the difference between the average dynamic and equilibrium capillary pressures increases. At any given node, the difference between the dynamic and equilibrium capillary pressure would always be zero because we assume local capillary equilibrium.

Similar effects have been observed in pore-network simulations (Gielen, 2003).

In order to show that our results are not affected by numerical discretisation, we have determined and compared the functions $\tau\left(\overline{S_w}\right)$ of two simulations with the same domain size (height $= 1.0$ m) but with different grid spacings (see Figure 12). The grid size has a small influence on the function $\tau(\overline{S_w})$ when the water saturation approaches $\overline{S_w} \rightarrow 1.0$. The discretisation needs to be fine enough to resolve the front correctly and to minimise numerical dispersion, thus a finer resolution is needed when calculating $\tau(\overline{S_w} \rightarrow 1.0)$. But, on the whole, it is clear that even for the $1\,\text{m} \times 1\,\text{m}$ domain, the discretisation is fine enough and our results are hardly influenced by numerical dispersion.

7. Influence of Heterogeneity

One presumption is that the dynamic effects as captured in the extended P_c–S_w relationship can be caused by soil heterogeneities in intrinsic

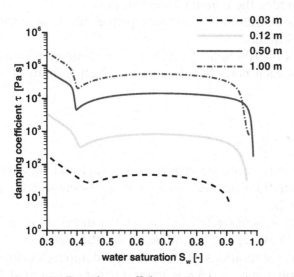

Figure 11. Damping coefficient τ (on log scale) as a function of the water saturation for three different domain sizes of the homogeneous coarse sand with the same initial pressure gradient of $\nabla P = 8.33 \times 10^3$ Pa/m ($P_c^B = 4.0 \times 10^3$ Pa).

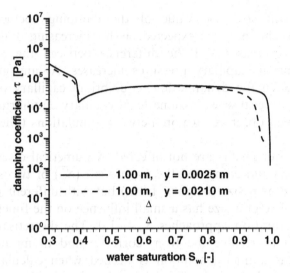

Figure 12. Damping coefficient τ (on log scale) for same domain size and different grid spacing.

permeabilities or constitutive properties. In order to test this assumption, a spatially-correlated isotopic random field was generated (see Section 3.3) for a 1 m × 1 m domain. Three numerical experiments with the same initial and boundary conditions (see Table I) were performed.

- Case 1: homogeneous set-up as reference case based on the parameters as given in Table II under the column "random field".
- Case 2: heterogeneous distribution of intrinsic permeabilities and homogeneous P_c–S_w relationship.
- Case 3: heterogeneous distribution of intrinsic permeabilities with entry pressure P_d scaled for each node based on the Leverett formula (Equation 7.1).

$$P_{d,i} = P_d^{ref} \cdot \sqrt{K^{ref}/k_i}. \tag{7.1}$$

Here $P_{d,i}$ denotes the entry pressure at node i. Porosity and the relative permeability-saturation relationship are assumed to be homogeneous and the same for all three cases.

Compared to the homogeneous reference case 1 the inclusion of a heterogeneous permeability (case 2) has an influence on the function $\tau\left(\overline{S_w}\right)$ (see Figure 13). For water saturations of $\overline{S_w} < 0.6$ the damping coefficient τ is higher for case 2 and for $\overline{S_w} > 0.6\,\tau$ is lower for case 2 compared to case 1. Whereas the differences between the phase pressures and the static capillary pressure does not differ significantly for the two cases, the satura-

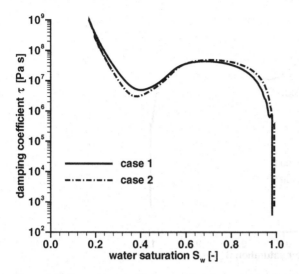

Figure 13. Damping coefficient τ (on log scale) as a function of water saturation for case 1 (homogeneous) and case 2 (heterogeneous permeability distribution) for a 1 m × 1 m domain.

tion rate of case 2 is up to 50% smaller than the one of case 1. The lower saturation rate would be caused by the regions of low permeabilities (see Figure 4) especially near the boundaries. The values for the intrinsic permeability differ by a factor of about 50. In order to eliminate the possibility that the shown influence on τ is only caused by the location of the low permeabilities with respect to the boundaries e.g. a Monte–Carlo simulation is required.

If a heterogeneous distribution of the entry pressure is generated based on the permeability field the functions $\tau(\overline{S_w})$ do not differ for large water saturations (see Figure 14). Notice that the maximum entry pressure $P_d =$ 1286 Pa of the scaled entry pressure distribution is still lower than the boundary capillary pressure. Consequently also in case 3 the entry pressure is overcome during the first time steps. An influence on the function $\tau(\overline{S_w})$ can be discerned for water saturations smaller than $\overline{S_w} < 0.5$. Here the values of the function $\tau(\overline{S_w})$ for case 3 are lower than the ones for case 2. However, the influence is not pro-nounced.

8. Influence of the Intrinsic Permeability and the BC-Parameter entry Pressure P_d

According to Stauffer's formula (see Equation (2.3)) the value of τ should be inversely proportional to intrinsic permeability and directly proportional to the square of the BC parameter entry pressure P_d.

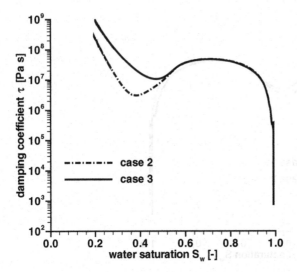

Figure 14. Damping coefficient τ (on log scale) as a function of water saturation for case 2 (heterogeneous permeability distribution) and case 3 (heterogeneous permeability and scaled entry pressure distribution) for a $1\,\text{m} \times 1\,\text{m}$ domain.

Here the value for the entry pressure was varied by 100%. In the first case we apply the heterogeneous permeability distribution described in Section 7 and a homogeneous P_c–S_w with an entry pressure of $P_d = 5.0 \times 10^2\,\text{Pa}$. In the second case the entry pressured is increased to $P_d = 1.0 \times 10^3\,\text{Pa}$.

For water saturations $\bar{S}_w > 0.75$ the resulting values of $\tau(\bar{S}_w)$ differ only slightly (see Figure 15). For decreasing water saturations the values of τ for the numerical experiment with the smaller entry pressure of $P_d = 5.0 \times 10^2\,\text{Pa}$ exceeds those of the higher entry pressure ($P_d = 1.0 \times 10^3\,\text{Pa}$) which obviously contradicts the functional relationship proposed by Stauffer. What is more, the damping coefficient τ for the numerical experiment applying an entry pressure of $P_d = 1.0 \times 10^3\,\text{Pa}$ changes sign. As τ was determined after Equation (6.1) it should be positive for drainage processes. However, at water saturations $0.23 < \bar{S}_w < 0.45$, the difference between the phase pressures and the equilibrium capillary pressure becomes negative.

Additionally, we will compare $\tau(\bar{S}_w)$ of case 1 (homogeneous) to the curve $\tau(\bar{S}_w)$ of the $1\,\text{m} \times 1\,\text{m}$ domain shown in Figure 11. The domain sizes of these two numerical experiments are the same. The entry pressure P_d of case 1 is 43% higher than in the case of the coarse sand used for the results shown in Figure 11 and the intrinsic permeability is reduced by a factor of nearly 1000. We find that the value of τ at a water saturation $\bar{S}_w = 0.6$ is increased by a factor of 1000, which suggests that the value of τ is inversely

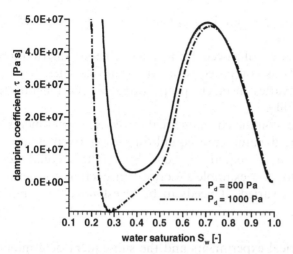

Figure 15. Damping coefficient τ as a function of water saturation two different entry pressures (random distribution of intrinsic permeabilities for a $1\,\text{m} \times 1\,\text{m}$ domain ($P_c^B = 1.0 \times 10^4\,\text{Pa}$).

proportional to the intrinsic permeability. With the function $\tau(S_w)$ of case 3 (see Figure 14) we have shown that the entry pressure only has an influence for small water saturations. Consequently, it stands to reason that on the macro-scale the influence of the intrinsic permeability on $\tau\overline{(S_w)}$ is more significant than the influence of the static capillary pressure-saturation relationship applied at the local scale.

9. Summary

In many two-phase flow situations, high pressure gradients can occur where the flow processes are far from the equilibrium conditions employed to determine the relationship between water saturation and capillary pressure. Some authors have proposed extended functional relationships to account for the varying flow conditions (see Section 2). Here the approach of Hassanizadeh and Gray (1993), an extended capillary pressure-saturation relationship (Equation (2.1)) has been analysed for the macro-scale based on numerical drainage experiments with the following results:

- The numerically determined values for τ may be as large as $\tau \approx 10^7\,\text{Pa s}$. Assuming the linear relationship holds and forcing it to run through the origin, τ is identified as a (non-linear) function of the water saturation on the macro-scale.
- The extended, linear P_c–S_w relationship (Equation (2.1)) does not hold over the full range of saturation rates.

- The boundary conditions (boundary capillary pressure) show an influence on the function $\tau(S_w)$.
- The values of $\tau(S_w)$ are influenced by a heterogeneous distribution of intrinsic permeabilities as compared to a homogeneous one.
- The magnitude of τ varies inversely proportional to the magnitude of the intrinsic permeability.
- During some of the conducted numerical drainage experiments the difference between the dynamic and equilibrium capillary pressures was not always larger than (or equal to) zero. So far, this could not be explained especially for the example described in Section 8 where only the entry pressure was varied. It needs to be investigated why this negative difference occurs.

The conducted numerical experiments and the subsequent determination of the damping coefficient τ revealed varying magnitudes of τ depending on the boundary conditions, the domain size and the soil parameters. For larger domain sizes, larger values of τ are obtained. That is, τ seems to be scale-dependent.

Acknowledgements

This research has been carried during a Visiting Research Fellowship for S. Manthey provided by the Delft University of Technology. The research was part of the TRIAS project "Upscaling microheterogeneities in two-phase flow in porous media" Delft Cluster Project no. 5.3.1. In Germany the research was funded by the German Research Foundation in the framework of the project MUSKAT (multi-scale-transport, He-2531/2-2). We greatly acknowledge the critical comments by two anonymous referees that helped to improve the paper.

References

Ataie-Ashtiani, B., Hassanizadeh, S. M., Oostrom, M., Celia, M. A. and White, M. D.: 2001, Effective parameters for two-phase flow in a porous medium with periodic heterogeneities. *J. Contam. Hydrol.* **49**: 87–100.
Barenblatt, G. I., Patzek, T. W. and Silin, D. B.: 2002, The mathematical model of non-equilibrium effects in water-oil displacement. In proceedings of SPE/DOE 13th Symposium on improved oil recovery. SPE 75169, Tulsa OK, USA.
Bastian, P.: 1999, Numerical computation of multiphase flows in porous media. Habilitation thesis. Christian-Albrechts-Universität Kiel, Germany.
DiCarlo, D. A. and Blunt, M. J.: 2000, Determination of finger shape using the dynamic capillary pressure. *Water Resour. Res.* **36**(9), 2781–2785.

Bourgeat, A. and Panfilov, M.: 1998, Effective two-phase flow through highly heterogeneous porous media: capillary non-equilibrium effects. *Comput. Geosci.* **2**, 191–215.

Brooks, R. H. and Corey, A. T.: 1964, Hydraulic properties of porous media. Hydrology Papers. Colorado State University.

Burdine, N. T.: 1953, Relative permeability calculations from pore size distribution data. Petroleum Trans. **198**, 71–77.

Dahle, H., Celia, M. A. and Hassanizadeh, S. M.: 2003, A bundle of capillary tubes model to investigate dynamic effects in the capillary pressure- saturation relationship. European Science Foundation (ESF) Exploratory Workshop: Recent advances in multiphase flow and transport in porous media, Delft University of Technology, Delft, The Netherlands, June 23–25.

Dykaar, B. B. and Kitanidis, P.: 1992, Determination of effective hydraulic conductivity for heterogeneous porous media using a numerical spectral approach: 1. *Meth. Water Resour. Res.* **28**(4), 1155–1166.

Gielen, T., Hassanizadeh, S. M., Nordhaug, H. and Leijnse, A.: 2003, Dynamic effects in multiphase flow: a pore-scale network approach. European Science Foundation (ESF) Exploratory Workshop: Recent advances in multiphase flow and transport in porous media, Delft University of Technology, Delft, The Netherlands, June 23–25.

Gielen, T., Hassanizadeh, S. M., Celia, M. A., Dahle, H. and Leijnse, A.: 2004, A pore-network approach to investigate dynamic effects in multiphase flow. Conference Proceedings of the Conference on Computational Methods in Water Resources, June 13th to 17th, 2004 in Chapel Hill, North Carolina, U.S.A., Vol. 1: 83–94.

Hassanizadeh, S. M. and Gray, W. G.: 1993, Thermodynamic basis of capillary pressure in porous media. *Water Resour. Res.* **29**(10), 3389–3405.

Hassanizadeh S. M., Celia, M. A. and Dahle, H. K.: 2002, Dynamic effect in the capillary pressure – saturation relationship and its impact on unsaturated flow. *Vadose Zone Hydrol.* **1**, 38–57.

Helmig, R.: 1997, *Multiphase Flow and Transport Processes in the Subsurface*. Springer, Berlin, Heidelberg.

Hilfer, R.: 1996, Transport and Relaxation Phenomena in Porous Media. Advances in Chemical Physics, XCII: 299–424.

Huber, R. U. and Helmig, R.: 1999, Multiphase Flow in Heterogeneous Porous Media: A Classical Finite Element Method Versus an IMPES-based Mixed FE/FV Approach. *Int. J. Num. Meth. Flu.* **1**(29), 899–920.

Kalaydijan, F. J.-M.: Dynamic capillary pressure curve for water/oil displacement in porous media: theory vs. experiment. SPE Conference. Washington D.C. 4–7 Oct. 1992. Paper no. 24183, p. 491–506.

Silin, D. and Patzek, T.: 2004, On Barenblatt's model of spontaneous countercurrent imbibition. *Transport in Porous Media* **54**, 297–322.

Singh, M. and Mohanty, K. K.: 2003, Dynamic modeling of drainage through three-dimensional porous materials. *Chem. Eng. Sci.*, **58**, 1–18.

Stauffer, F.: 1978, Time dependence of the relationship between capillary pressure, water content and conductivity during drainage of porous media. In proceedings of the IAHR conference on scale effects in porous media, Thessaloniki, Greece.

Topp, G. C., Klute, A. and Peters, D. B.: 1967, Comparison of water content-pressure head obtained by equilibrium, steady-state and unsteady state methods. *Soil. Sci. Am. Proc.* **31**, 312–314.

Transp Porous Med (2005) 58:147–172
DOI 10.1007/s11242-004-5473-5

Dynamic Capillary Pressure Mechanism for Instability in Gravity-Driven Flows; Review and Extension to Very Dry Conditions

JOHN L. NIEBER[1,*], RAFAIL Z. DAUTOV[2], ANDREY G. EGOROV[3] and ALEKSEY Y. SHESHUKOV[4]

[1]*Department of Biosystems and Agricultural Engineering, University of Minnesota, 1390 Eckles Ave, St. Paul, MN 55108, U.S.A.*
[2]*Faculty of Computational Mathematics and Cybernetics, Kazan State University, Kremlyovskaya 18, Kazan 420008, Russia*
[3]*Chebotarev Research Institute of Mathematics and Mechanics, Kazan State University, Universitetskaya 17, Kazan 420008, Russia*
[4]*Department of Biosystems and Agricultural Engineering, University of Minnesota, 1390 Eckles Ave, St. Paul, MN 55108, U.S.A.*

(Received 10 September 2003; in final form: 24 March 2004)

Abstract. Several alternative mathematical models for describing water flow in unsaturated porous media are presented. These models are based on an equation for conservation of mass of water, and a generalized linear law for water flux (Darcy's law) containing a term called the dynamic capillary pressure. The distinct form of each alternative model is based on the specific form of expression used to describe the dynamic capillary pressure. The conventional representation arises when this pressure is set equal to the equilibrium pressure given by the capillary pressure – saturation function for unsaturated porous media, and this conventional approach leads to the Richards equation. Other models are derived by representing the dynamic capillary pressure by a rheological relationship stating that the pressure is not given directly by the capillary pressure – saturation function. Two forms of rheological relationship are considered in this manuscript, a very general non-equilibrium relation, and a more specific relation expressed by a first-order kinetic equation referred to as a relaxation relation. For the general non-equilibrium relation the system of governing equations is called the general Non-Equilibrium Richards Equation (NERE), and for the case of the relaxation relation the system is called the Relaxation Non-Equilibrium Richards Equation (RNERE). Each of the alternative models was analyzed for flow characteristics under gravity-dominant conditions by using a traveling wave transformation for the model equations, and more importantly the flow described by each model was analyzed for linear stability. It is shown that when a flow field is perturbed by infinitesimal disturbances, the RE is unconditionally stable, while both the NERE and the RNERE are conditionally stable. The stability analysis for the NERE was limited to disturbances in the very low frequency range because of the general form of the NERE model. This analysis resulted in what we call a low-frequency criterion (LFC) for stability.

---

*Author for correspondence: e-mail: nieber@umn.edu

This LFC is also shown to apply to the stability of the RE and the RNERE. The LFC is applied to stability analysis of the RNERE model for conditions of initial saturation less than residual.

1. Introduction

The instability of unsaturated flows during infiltration or redistribution of water within soils and the vadose zone has been identified to be one form of preferential flow through which fast transport of contaminants might reach ground water resources (Glass et al., 1988; Nieber, 2001). Due to the recognition of the importance of this process, much effort has been expended in the experimental and mathematical analyses of gravity-driven unstable flows in unsaturated porous media with the idea that the development of a complete theory and parameterization of unstable flows should provide important components of soil hydrology and solute transport models.

The importance of gravity-driven unstable flows in unsaturated soils was first recognized upon the publication of the first definitive study of gravity-driven fingering in layered porous media in the paper by Hill and Parlange (1972). Several earlier reports of gravity-driven fingering had been reported by other investigators (e.g. Tabuchi, 1961; Smith, 1967) but those studies were not definitive enough with relation to gravity-driven fingering to capture the interest of soil physics and soil hydrology researchers. The work of Hill and Parlange did indeed capture that interest and motivated extensive experimental and theoretical work. A number of experimental studies of gravity-driven fingering followed, using two-dimensional slab chambers filled with porous media packed either as two homogeneous layers of different texture (Diment and Watson, 1985; Glass et al., 1989a,c; Baker and Hillel, 1990; Wang et al., 1998a, b), as completely homogeneous systems (Selker et al., 1992a; Liu et al., 1994b; Bauters et al., 2000; Deinert et al., 2002), or as heterogeneous systems (Sililo and Tellam, 2000). The first definitive experimental study for fingering in field soils was given by Starr et al. (1978).

Experimental data quantifying unstable flow have been derived using visual observations of finger width and velocity, flow within individual fingers, water pressure measurements within individual fingers (Selker et al., 1992b), and water saturation distributions within fingers using either light transmission (Glass et al., 1989a), gamma-ray attenuation (Bauters et al., 2000), or neutron radiography (Deinert et al., 2002). One point that comes out of all of these studies is that fingers are observed to occur when the initial saturation is below the residual. For conditions where the initial saturation is above residual, fingers of width less than the width of the experimental chamber are not observed to occur, at least by the methods of observation utilized to date.

The first mathematical analysis of gravity-driven unstable flow was formulated by Raats (1973) wherein he used the Green-Ampt model of infiltration, a sharp front model, as the basis for his analysis. The same basic model was used in the linear stability analysis presented by Philip (1975a, b) for gravity-driven flows. Philip's analysis was similar to that presented by Saffman and Taylor (1958) and Chuoke et al. (1959) for viscous fingering, and showed that flows become unstable when the pressure gradient opposes the flow. His spectral analysis provided expressions to calculate the critical perturbation wavelength and estimates of finger widths, but the derived expressions contained properties related to Hele-Shaw cells and not to real soils. Philip (1975a) explained that the Green-Ampt approximation has significant limitations as the basis for the stability analysis of infiltrating flows in real soils, and argued the need to perform analysis of flow instability for the conditions where the wetting front is not sharp. As explained by Philip (1975a) this would mean a stability analysis of the Richards equation, which he stated would prove to be difficult. This need was partially met by the work of Parlange and Hill (1976) wherein the stability analysis of the Green-Ampt type of model was extended to real soils by imposing a diffuse structure to the wetting front. Their analysis provided expressions for estimates of finger width as a function of imposed flow, saturated hydraulic conductivity of the soil, initial moisture content, and sorptivity. The analysis of Parlange and Hill has had a lasting impact as many formulae for estimating finger size have been derived based on their original work (Wang et al., 1989a; Glass et al., 1989b; Liu et al., 1994a; deRooij and Cho, 1999).

The first stability analysis to the full Richards equation was presented by Diment et al. (1982). In their analysis the basic equation of flow was given by the Richards equation, and the pressure for the flow field was perturbed. The form of their resulting perturbation equation was not tractable to analytical solution, so a numerical solution was sought instead and results were reported by Diment and Watson (1983). For the limited cases considered they concluded that flows governed by the Richards equation are stable to infinitesimal perturbations. But their results were based on a numerical solution and therefore it was not possible to provide a general result for all conditions wherein linear stability analysis would apply.

Kapoor (1996) presented an analytical solution to the perturbed steady-state Richards equation and concluded that steady-state flows governed by the Richards equation are unconditionally stable for the exponential form of the hydraulic conductivity – pressure function and conditionally stable for other forms such as the Brooks and Corey (1964) and van Genuchten (1980) forms of the hydraulic conductivity – pressure function. Ursino (2000) performed a similar analysis to that of Kapoor except in her case time-dependent flows were considered. She showed that for the exponential form of the hydraulic conductivity – pressure function, Richards' equation is unconditionally stable, and therefore concluded that flow instabilities must originate from some pore-scale process not included in the conventional upscaling of the flow equation.

In the most recent published analyses the saturation flow field governed by the saturation form of the Richards equation has been subjected to stability analysis by Du et al. (2001) and Egorov et al. (2002, 2003). The resulting perturbed flow equation was evaluated analytically by Du et al. but due to it complex nature their analysis was incomplete, leading them to the conclusion that the Richards equation is conditionally stable. One important feature of the work by Egorov et al. was that it provided a complete analytical result for the perturbed flow equation and the results led to the conclusion that the Richards equation is unconditionally stable. Their result is consistent with the nonlinear stability analysis given by Otto (1996, 1997), which concluded that the Richards equation is unconditionally stable to all perturbations (infinitesimal and finite) in homogeneous unsaturated porous media. Egorov et al. (2003) extended the nonlinear analysis of Otto by showing that the Richards equation is unconditionally stable to all perturbations even for heterogeneous porous media.

These final results point to the fact that flow instabilities that occur in gravity-driven flows must result from a flow process not included (explicitly or implicitly) in the Richards equation, and therefore we must conclude that the flow process is not described adequately by the conventional Darcy law. In line with the conclusion of Ursino (2000), we conclude that the process that causes flows to become unstable must arise from pore-scale phenomena not included in the conventional governing equations. Going further and following the work of Hassanizadeh and Gray (1993) we would postulate that one possible pore-scale process that could cause instabilities is the process described by dynamic capillary pressure – saturation relations. The results presented to date in Egorov et al. (2002, 2003) seem to support this postulate.

This manuscript will provide an overview of the current state of understanding from a mathematical analyses standpoint, of gravity-driven flow instabilities in unsaturated porous media. The presentation will review the results presented to date by Egorov et al. (2002, 2003) with respect to the unconditional stability of the Richards equation, and the conditional stability of flows described by models that include the dynamic capillary pressure effect. In addition, we will present some new results that relate to the extension of the models of Egorov et al. to the (dry) range of saturations below residual saturation.

2. Overview of Selected Governing Equations for Unsaturated Flow

The three-dimensional mass balance equation is written in non-dimensional form as

$$\frac{\partial s}{\partial t} - \nabla \cdot (K(s)\nabla p) - \frac{\partial K(s)}{\partial z} = 0, \tag{1}$$

where s is the effective saturation equal to $(S - S_r)/(1 - S_r)$, S is the water saturation, S_r is the residual saturation, p is the water pressure, $K(s)$ is the unsaturated hydraulic conductivity, and z is the z-coordinate taken positive upward opposite to the direction of gravity, and t is the time. This equation is based on the substitution of Darcy's law into the equation for conservation of mass. Different forms of the mass balance equation can be obtained depending on the form of the pressure function p. The forms available to define the pressure function are numerous. The first one is based on a conventional formulation and thereby leads to the Richards Equation (RE) and is given by

$$p = P(s). \tag{2}$$

This equation describes the conventional equilibrium relation between water pressure and water saturation, and can be non-hysteretic or hysteretic. The substitution of p from relation (2) into equation (1) yields the conventional RE given by

$$\frac{\partial s}{\partial t} - \nabla \cdot (K(s)\nabla P(s)) - \frac{\partial K(s)}{\partial z} = 0 \tag{3}$$

or in saturation form the equation is

$$\frac{\partial s}{\partial t} - \nabla \cdot (D(s)\nabla s) - \frac{\partial K(s)}{\partial z} = 0, \tag{4}$$

where $D(s) = K(s)P'(s)$. Substitution of relation (2) into Darcy's law ($q = -K(s)\nabla p - K(s)e_z$) gives

$$q = -K(s)\nabla P(s) - K(s)e_z, \tag{5}$$

where e_z is the unit vector in the vertical.

A more general form of the pressure function is given by the relation

$$F\left(s, p, \frac{\partial s}{\partial t}, \frac{\partial p}{\partial t}, \dots\right) = 0. \tag{6}$$

This equation indicates a non-equilibrium relation between water pressure and saturation. The dependence of the pressure on the saturation and temporal derivatives of the saturation and the pressure are indicated by the terms included in the argument of the function. The idea for non-equilibrium relations for water pressure and saturation is based on experimental evidence presented in various studies (Kirkham and Feng, 1949; Nielsen et al., 1962; Rawlins and Gardner, 1963; Topp et al., 1967; Smiles et al., 1971; Wildenschild et al., 2001) and theoretical considerations (Hassanizadeh and Gray, 1993; Hassanizadeh et al., 2002; Dahle et al., 2002). Since this function has a generalized form it is not possible to introduce it directly into the mass balance equation and thereby provide a distinct equation. But an analysis of Equation (1) coupled with relation (6) is possible as will be shown in Section 3. The combination of equations (1) and (6) will hereafter be referred to as the NERE model.

A specific form of relation (6) that we will spend a significant part of the next sections describing is given by the relaxation equation

$$\tau(s, p)\frac{\partial s}{\partial t} = p - P(s), \tag{7}$$

where $\tau(s, p)$ is a relaxation parameter in the kinetic rheological relation for non-equilibrium capillary pressure – saturation relations. Substitution of function p from relation (7) into Equation (1) leads to

$$\frac{\partial s}{\partial t} - \nabla \cdot (K(s)\nabla P(s)) - \nabla \cdot \left(K(s)\nabla \left(\tau(s, p)\frac{\partial s}{\partial t} \right) \right) - \frac{\partial K(s)}{\partial z} = 0 \tag{8}$$

and into Darcy's law we get

$$q = -K(s)\nabla P(s) - K(s)\nabla \left(\tau(s, p)\frac{\partial s}{\partial t} \right) - K(s)e_z. \tag{9}$$

The equations given by equation (1) and relation (7) or alternatively Equation (8) will hereafter be referred to as the RNERE model. In relation (9) we have the usual gradient of the equilibrium pressure term, but also the gradient of the relaxation term which contains the temporal rate of change of saturation.

For most of the analyses to follow we consider conditions where the saturation falls in the range $0 < s < 1$, however, in Section 4.2.2 we consider the case where the initial saturation is less than residual.

3. Review of Stability Analyses of the RE

Linear stability analyses of the RE have been presented by Diment and Watson (1983), Ursino (2000), Kapoor (1996), Du et al. (2001) and Egorov et al. (2002, 2003). Diment and Watson, Ursino, and Kapoor all started with the pressure-based form of the RE, while Du et al. and Egorov et al. utilized the saturation-based form of the RE.

Starting with the saturation-based form of the RE the traveling wave equation is derived using the transformation variables

$$s = s(\xi), \quad \xi = z + Vt \tag{10}$$

subject to the boundary conditions

$$s(-\infty) = s_-, \quad s(+\infty) = s_+, \quad 0 < s_- < s_+ < 1, \tag{11}$$

where the velocity V of the traveling wave is given by

$$V = \frac{K(s_-) - K(s_+)}{s_- - s_+}. \tag{12}$$

Applying these variables to equation (4) yields the traveling wave form of the RE,

$$V\frac{ds}{d\xi} - \frac{d}{d\xi}\left(D(s)\frac{ds}{d\xi}\right) - \frac{dK(s)}{d\xi} = 0. \tag{13}$$

The solution for equation (13) subject to the boundary conditions at infinity ($\xi = +\infty$) is given as (Philip, 1957)

$$\xi(s) - \xi_* = \int_{s_*}^{s} \frac{D(s)ds}{v(s - s_+) - K(s) + K(s_+)}, \tag{14}$$

where ξ_* is the coordinate location of the arbitrarily selected saturation s_* ($s_- < s_* < s_+$). The inverse of the function $\xi(s)$ is called the basic solution and will be designated as $s_0(\xi)$. A typical plot of this solution is presented in Figure 1. By the nature of the solution and the included functions ($K(s)$ and $P(s)$) the saturation decreases monotonically from s_+ at $+\infty$ to s_- at $-\infty$.

The stability analysis is based on a small perturbation applied to the basic solution $s_0(\xi)$. The perturbed saturation field is represented as

$$s(x, y, z, t) = s_0(\xi) + \varepsilon e^{i\omega_x x + i\omega_y y + kt} s_1(\xi) + O(\varepsilon^2), \tag{15}$$

where the ω_x and the ω_y are characteristic wave numbers in the x and y directions, respectively, the k is the amplification factor (or growth factor), the function $s_1(\xi)$ describes the variation of the bounded perturbation in the ξ coordinate and vanishes at $\pm\infty$, and ε scales the magnitude of the perturbation.

The perturbed solution is obtained by substituting expression (15) into equation (4) and dropping terms of order ε^2. Accounting for the equation for the basic solution and collecting like terms we arrive at a locally

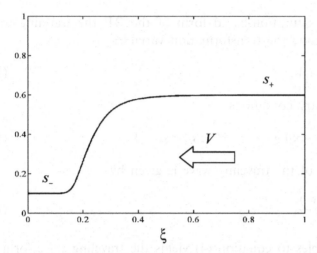

Figure 1. A typical plot of saturation *s* versus dimensionless variable ξ as a result of the solution of the traveling wave form of the RE. This solution comprises the basic solution s_0.

linearized perturbation equation in the form of a spectral problem for determining nontrivial s_1 and $k = k(\omega)$ for arbitrary ω, that is

$$-\frac{\mathrm{d}^2}{\mathrm{d}\xi^2}(D(s_0)s_1) + \frac{\mathrm{d}}{\mathrm{d}\xi}\left((V - K'(s_0))s_1\right) + \omega^2 D(s_0)s_1 = -ks_1, \quad -\infty < \xi < \infty.$$

(16)

An equation of similar form to equation (16) was also derived by Du *et al.* (2001).

Now the problem is the analysis of the spectrum of the problem (16). If the equation admits a solution (s_1, k) with $k > 0$ then we have instability of the RE. If all solutions have $k < 0$ then the flow governed by the RE is stable. Analytical investigation of the spectral problem is difficult in the form of equation (16) because the equation is not self-adjoint. To make the analytical study tractable we perform a transformation using new variables ζ and θ to replace ξ and s_1 respectively,

$$\zeta = \int \frac{\mathrm{d}\xi}{\sqrt{D(s_0)}}, \quad \theta = \frac{D^{1/4}(s_0)}{\sqrt{s_0'}}s_1,$$

(17)

which leads to

$$-\frac{\mathrm{d}^2\theta}{\mathrm{d}\zeta^2} + (\omega^2 D + F)\theta = -k\theta, \quad -\infty < \zeta < \infty,$$

(18)

where

$$F = \frac{1}{B}\frac{\mathrm{d}^2 B}{\mathrm{d}\zeta^2}, \quad B = D^{1/4}\sqrt{s_\mathrm{o}'}.$$

Equation (18) is self-adjoint and therefore has been made tractable to theoretical analysis.

In mathematical physics equation (18) is known as the Schrödinger equation, and for different forms of the potential $(\omega^2 D + F)$ the spectral problem for the Schrödinger equation has been studied (Carmona and Lacroix, 1999), and therefore known techniques can be used to investigate the problem presented by equation (18). Such an analysis was performed in Egorov *et al.* (2003) in which it was shown that the spectrum of the problem is negative (i.e., $k < 0$) for all non-zero frequencies ω of the perturbation.

So we can now state that the RE is unconditionally stable to infinitesimal perturbation. But it is legitimate to ask whether the occurrence of finite perturbations to the basic flow would be stable. This query has been addressed by Otto (1996, 1997) and by Egorov *et al.* (2003). It was shown by Otto that the type of equation given by the RE is unconditionally stable for perturbations of finite magnitude in the case of homogeneous porous media. Otto's work was extended by Egorov *et al.* to show that the same conclusion is found for the case of heterogeneous media. While the linear stability analysis outlined in the foregoing description points toward unconditionally stability of the RE it does not prove the stability of the RE for all conditions. But the nonlinear stability analyses of Otto and Egorov *et al.* provides strong proof that the RE is unconditionally stable for all conditions. This also means that upscaling of the RE over a heterogeneous domain will lead to a governing equation that possesses the property of unconditional stability.

4. Stability Analyses of Selected Nonequilibrium Models

In this section we will consider the two types of nonequilibrium models that were outlined in Section 2, that is the NERE and the RNERE. The NERE will be described under the heading of the general nonequilibrium model, while the RNERE will be described under the heading of the relaxation nonequilibrium model.

4.1. GENERAL NONEQUILIBRIUM MODEL

The general nonequilibrium model was presented in Section 2 by the coupling of equation (1) and relation (6) and referred to as the NERE. Due to the non-specific form of relation (6) it is not possible to derive an explicit equation for analysis, but as shown by Egorov *et al.* (2003) it is possible to derive a useful stability criterion for the NERE.

A linear stability analysis is performed on the NERE using the same procedure we used for the stability analysis of the RE. First we define a basic solution based on the traveling wave form of the NERE. The traveling wave form of equation (1) for that model is independent of the form of the rheological relation. That traveling wave equation is given by

$$V\frac{ds}{d\xi} - \frac{d}{d\xi}\left(K(s)\frac{dp}{d\xi}\right) - \frac{dK(s)}{d\xi} = 0. \tag{19}$$

After integrating once and applying the boundary conditions we have the following result,

$$\frac{dp}{d\xi} = \frac{V(s - s_+) + K(s_+) - K(s)}{K(s)}. \tag{20}$$

The transformation to the traveling wave variable for relation (6) gives

$$F\left(s, p, V\frac{ds}{d\xi}, V\frac{dp}{d\xi}, \ldots\right) = 0. \tag{21}$$

Let us suppose that the solution to the traveling wave equation (20) and relation (21) exists and are represented by $s_0(\xi)$ and $p_0(\xi)$, respectively. The exact form of these basic solutions will depend on the form of the rheological relation (6), but it is not necessary to specify the form of that relation to perform the following stability analysis.

The next step in the linear stability analysis is to superimpose three-dimensional perturbations in saturation and pressure onto the basic solution in the form

$$s(x, y, z, t) = s_0(\xi) + \varepsilon e^{i\omega_x x + i\omega_y y + kt} s_1(\xi) + O(\varepsilon^2), \tag{22}$$

$$p(x, y, z, t) = p_0(\xi) + \varepsilon e^{i\omega_x x + i\omega_y y + kt} p_1(\xi) + O(\varepsilon^2). \tag{23}$$

These perturbation expressions are substituted into equation (1) and relation (6), and terms of order ε^2 are dropped to yield a system of locally linearized perturbation equations in the form of a spectral problem for determining nontrivial p_1 and s_1, corresponding to spectral parameter k for arbitrary frequencies ω. This spectral problem is expressed by

$$\frac{dA}{d\xi} + \omega^2 K(s_0) p_1 = -ks_1, \tag{24}$$

$$\Phi\left(s_1, p_1, \frac{ds_1}{d\xi}, \frac{dp_1}{d\xi} \ldots ; s_0, p_0, \frac{ds_0}{d\xi}, \frac{dp_0}{d\xi}, \ldots ; k\right) = 0, \tag{25}$$

where A is the flux perturbation given by

$$A = -K(s_0)\frac{dp_1}{d\xi} - K'(s_0)\left(1 + \frac{dp_0}{d\xi}\right)s_1 + Vs_1. \tag{26}$$

Integration of equation (24) with the boundary condition such that A vanishes as $\xi \to \pm\infty$, leads to the integrals

$$\omega^2 \int_{-\infty}^{+\infty} K(s_0)p_1 d\xi = -k \int_{-\infty}^{+\infty} s_1 d\xi. \tag{27}$$

It is not possible to evaluate the signs of the integrals in equation (27) and thereby be able to find the sign of $k = k(\omega)$ without knowing the specific forms for s_0, s_1 and p_1. However, it is possible to perform an asymptotic analysis to derive an asymptotic solution for the eigenvalue k_0 at low frequency($\omega \ll 1$). This analysis begins by establishing the fact that the eigenfunction for s_1 and p_1 are equal to $ds_0/d\xi$ and $dp_0/d\xi$ when $\omega = 0$ as shown in Egorov et al. (2003). At low frequency the eigenvalue k_0 and the eigenfunctions s_1 and p_1 can be expanded in powers of ω^2 as

$$k_0 = 0 + b\omega^2 + \cdots,$$
$$s_1 = ds_0/d\xi + s_*\omega^2 + \cdots,$$
$$p_1 = dp_0/d\xi + p_*\omega^2 + \cdots$$

Substituting these expressions into equation (27) and dropping terms of order ω^4 leads to

$$b = -\frac{C}{s_+ - s_-}$$

with

$$C = \int_{-\infty}^{+\infty} K(s_0)\frac{dp_0}{d\xi}d\xi. \tag{28}$$

Therefore we have

$$k_0 = -\frac{C}{s_+ - s_-}\omega^2 + O(\omega^4). \tag{29}$$

Since ω^2 and $(s_+ - s_-)$ are inherently positive, the condition given by equation (29) means that flows will be unstable ($k_0 > 0$) when $C < 0$. Examining the expression for C in equation (28), the value of C will be negative when the pressure field is sufficiently non-monotonic. This result about the pressure gradient being opposed to the flow for gravity-driven unstable flows was also found by Raats (1973), Philip (1975a) and Parlange and Hill (1976).

The low frequency criterion established here will be used in later sections to evaluate the stability of the RNERE models presented in those

sections. But here we should end with a note about the use of the low frequency criterion as a tool to evaluate the stability of the RE model. It was
established in Section 3 that for infiltrating flows the pressure and saturation profiles in the RE model are monotonic, meaning that the pressure
gradient $dp_0/d\xi$ is positive for all cases of the RE model. Using this result
in equation (28) leads to $C > 0$, and from expression (29) the value of $k_0 < 0$
for all small ω, meaning that infiltrating flows governed by the RE will be
stable.

4.2. RELAXATION NONEQUILBRIUM MODEL

The relaxation non-equilibrium model is a special case of the generalized nonequilibrium model where the dynamic pressure is given by a first
order kinetic rheological relation as in relation (7). For this model we
have examined two conditions, one in which the initial saturation is above
the residual, and the other where the initial saturation is less than the
residual. These distinct cases are both important because most models of
unsaturated flow have involved saturation conditions above residual, while
in many realistic field conditions, and also in many laboratory experiment conditions the saturations within the flow domain can be below
residual. Each of these cases will be examined in the following subsec-
tions.

4.2.1. *RNERE Model for Initial Moisture Greater than Residual*
The RNERE model is given by the combination of equations (1) and relation (7). The traveling wave form of those equations are given by the coupled equations

$$\frac{ds}{d\xi} = \frac{p - P(s)}{V\tau(p, s)}, \tag{30}$$

$$\frac{dp}{d\xi} = \frac{V(s - s_+) + K(s_+) - K(s)}{K(s)} \tag{31}$$

subject to the initial conditions $s(-\infty) = s_-$ and $p(-\infty) = P(s_-)$. Equation
(31) was obtained after once integrating with respect to ξ and applying the
boundary condition $s = s_+$ at $\xi = \infty$.

We now investigate the situation where τ is factorized into a constant
τ_o, a function of pressure $\tau_p(p)$ and a function of saturation $\tau_S(s)$. This
factorization is expressed by

$$\tau(s, p) = \tau_0\tau_p(p)\tau_S(s). \tag{32}$$

A number of functions could be used to express the pressure and the saturation dependencies. We have used the following functions:

$$\tau_S(s) = \frac{dP(s)}{ds}, \tag{33}$$

$$\tau_S(s) = s^\gamma (1-s)^\delta, \tag{34}$$

$$\tau_p(p) = (p_* - p)^\eta, \quad \eta > 0. \tag{35}$$

The other parameters that need to be specified are those defining the capillary pressure – saturation relation, and the hydraulic conductivity – saturation relation. The more common types of forms used in modeling unsaturated flows are those referred to as the van Genuchten relations and the Brooks–Corey relations. Both of these types of relations were used in the present study. The van Genuchten relations (van Genuchten, 1980) are given by

$$P(s) = -(s^{-1/nm} - 1)^{1/n}, \tag{36}$$

$$K(s) = \sqrt{s}\left[1 - \left(1 - s^{1/m}\right)^m\right]^2, \tag{37}$$

where n and m are porous media dependent parameters. The relation for the hydraulic conductivity is for the special case of a Mualem–Burdine type of pore-scale model where $m = 1 - 1/n$.

The relations for $P(s)$ and $K(s)$ for the Brooks–Corey formulation (Brooks and Corey, 1964) are given by

$$P(s) = -s^{-\beta}, \quad 0 < \beta < 1, \tag{38}$$

$$K(s) = s^\alpha, \quad \alpha > 2. \tag{39}$$

For these models hysteresis in the $P(s)$ function was represented by the Mualem (1974) independent domain model.

4.2.1.1. Basic Solution. In this section we will present numerical solutions of equations (30) and (31) with the associated initial condition. A similar analysis was presented by Cuesta *et al.* (2000) wherein they analyzed the existence, uniqueness and monotonicity of the solution to Equation (8) instead of the system of equations (1) and (7). In the following we will show results from our own calculations about monotonicity of the solution of this system of equations, and where it is pertinent we will relate our results to those of Cuesta *et al.*

Solutions (that is basic solutions) to Equations (30) and (31) were performed numerically using the relaxation function expressed by relation (32) with relations (33) and (35) in particular. A sample set of solutions are illustrated in Figure 2, where all parameters are kept constant except for the value of τ_0. The invariant parameters are: $s_+ = 0.6$, $s_- = 0.1$, $n = 10$,

$\alpha = 2.5$, and $\eta = 0$. The value set for η makes the relaxation coefficient independent of pressure for these simulations. The results for six values of τ_0 are presented in the plot. It is seen that as the value of τ_0 increases the solution becomes more non-monotonic. For the case with $\tau_0 = 0$ the solution is the same as the solution to the RE, and the profile is seen to be monotonic.

The magnitude of the relaxation coefficient necessary to produce non-monotonicity in the basic solution is estimated from the parameter $\tau_F(s_+)$, given by,

$$\tau_F = \tau_F(s_+) = \frac{\left(s_+ P'(s_+)\right)^2}{4\left(s_+ K'(s_+) - K(s_+)\right)}. \tag{40}$$

Equation (40) is derived from analysis of the traveling wave equations (30) and (31) with the underlying assumption that $s_+ \gg s_-$. For the parameters used in producing Figure 2, equation (40) gives $\tau_F(s_+) = 0.036$. Therefore, when $\tau(s_+) \geqslant 0.036$ the traveling wave solution is non-monotonic. For the cases shown in Figure 2, this occurs for the values of τ_0 exceeding about 0.08.

It is observed that the tails of the non-monotonic profiles shown in Figure 2 are oscillatory. This feature can be understood upon viewing the

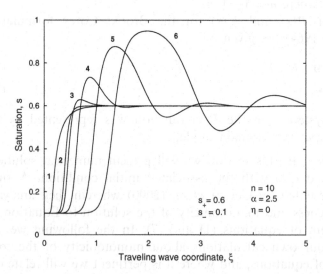

Figure 2. Plots of saturation versus traveling wave variable for various values of τ_0. The curve labels 1 to 6 correspond to the following values of $\tau_0 = 0$, 0.07, 0.1, 0.2, 0.5 and 1.0. The relaxation function $\tau(s, p)$ is given by relation (32) in conjunction with (33) and (35). The critical value of τ_F for non-monotonicity is 0.036. The first two curves are monotonic and $\tau(s_+, p_+) > \tau_F$, while curves 3–6 are non-monotonic and $\tau(s_+, p_+) > \tau_F$. The larger the value of τ, the more non-monotonic the saturation profile.

phase plane shown in Figure 3(a), which is derived from the same solution for $\tau_0 = 0.5$ shown in Figure 2. There we see that the trace starting at s_- ends at a focus point around s_+. In contrast, as shown in Figure 3(b) for the monotonic solution (derived with $\tau_0 = 0.07$) the end point at s_+ is nodal and therefore the tail is non-oscillatory. Although the result is not shown here, we found the tail of a non-monotonic solution to be non-oscillatory (ends at a nodal point at s_+) when the equilibrium capillary pressure – saturation relation $P(s)$ is hysteretic.

For the model just presented we used relation (33) which yields a saturation dependence to the relaxation coefficient similar in behavior to that derived by Panfilov (1998) in his analysis for upscaling dynamic capillary pressure for two-phase flows. This relation shows (see Figure 4) that the saturation dependent part of the coefficient is unbounded at the extreme ends of the effective saturation range. In using this model we found that as $s_- \to 0$, the pressure at the wetting front becomes unbounded and physically unrealistic. A sample result of this is shown in Figure 5(a). The parameters used to derive this were the same as those used to derive Figure 2, with $\tau_0 = 0.5$. It is observed that as s_- progressively decreases, the water pressure at the wetting front increases to the point where it becomes positive, which is a physically unrealistic result. Even further reduction in s_- leads to pressure at the front approaching infinity.

A solution we discovered to resolve this problem of unbounded pressure was to apply a non-unity pressure factorization into the relaxation coefficient, such as the factorization given by relation (35). With this factorization, as $s_- \to 0$ the pressure at the wetting front is limited to p_*. A sample result of this solution is shown in Figure 5(b), wherein the parameters are

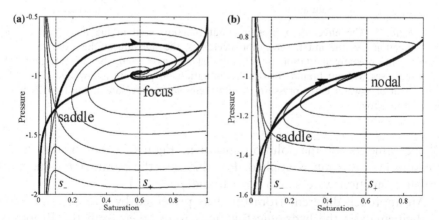

Figure 3. A typical phase plane plot for the basic solution to the RNERE equations for two cases of τ: (a). Where the τ is sufficiently large ($\tau \geqslant \tau_F$) to lead to a non-monotonic saturation profile, and (b). where the τ is small enough ($\tau \ll \tau_F$) to lead to a monotonic saturation profile.

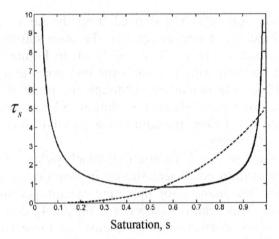

Figure 4. Alternative forms of the saturation component of the relaxation coefficient. These are based on Equations (33) (solid line) and (34) (dash line).

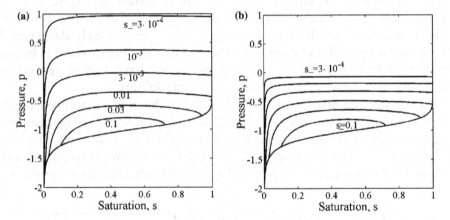

Figure 5. The effect of a pressure limit function used in the relaxation coefficient function. As the initial saturation decreases, the pressure at the front increases. In (a) the pressure limitation is not imposed and as $s_- \to 0$, the pressure at the front increases without bound. In (b) the limitation on pressure is imposed and is in the form of a water entry pressure p_* and therefore the pressure at the front is bounded from above.

all the same as for the solution shown in Figure 5a, but in this case the pressure factorization is applied. For this case the parameter values in the pressure function were set to $\eta = 1.0$ and $p_* = -0.1$.

An approach different from this was presented by Cuesta et al. (2000). In their analysis the hydraulic functions were taken from the Brooks–Corey functions. For the relaxation coefficient they used the formula $\tau = \tau_0 s^\gamma$ (same as relation (34) but with $\delta = 0$). A qualitative description of this function is illustrated in Figure 4 with $\gamma > 0$, where it is seen that the relaxation

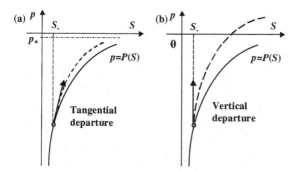

Figure 6. Phase plane diagrams showing the effect of porous media hydraulic and non-equilibrium characteristics on the shape of the phase plane trace. The dividing point between the two behaviors is the parameter $\gamma = (\alpha - 2 - 2\beta)$. For (a) the $\gamma > (\alpha - 2 - 2\beta)$, and the solution to the traveling wave equation is bounded from above. For (b) the $\gamma < (\alpha - 2 - 2\beta)$ and the traveling wave solution is not bounded, and therefore the solution does not exist.

coefficient approaches zero as the saturation approaches zero while it is finite and non-zero at full saturation.

Cuesta *et al.* (2000) stated that the solution to the traveling wave form of the RNERE exists and is unique for all intial conditions when $s_- > 0$. However, for the condition where $s_- \to 0$, they found the same result as we did where the pressure at the front became unbounded. Rather than use a pressure factorization as we did, they instead defined the range of parameters that would allow the solution pressure to be bounded as $s_- \to 0$. Their analysis showed that for the solution to be bounded the parameter γ needs to be constrained by the inequality $\gamma > (\alpha - 2 - 2\beta)$.

A qualitative result showing the effect of setting γ according to this criterion is illustrated in Figure 6 where the phase plane for saturation versus. pressure is plotted. Shown on each plot are the equilibrium pressure curve and the trace set by the evolution of saturation in the traveling wave solution. For the case shown in Figure 6(a) the value of $\gamma > (\alpha - 2 - 2\beta)$. For this case the trace makes a tangential departure from the equilibrium curve, and at the end the trace reaches a pressure that is bounded from above. In contrast, for the case shown in Figure 6(b), the value of $\gamma < (\alpha - 2 - 2\beta)$, and for this situation the trace makes a vertical departure from the phase plane, effectively causing the trace to reach pressures that are not bounded from above.

4.2.1.2. Stability Analysis for the RNERE. The linear stability analysis for the RNERE is performed similarly to that for the NERE (see equations (24) and (25)). We need only to specify the equation due to the relaxation law. The resulting perturbed equations are given by

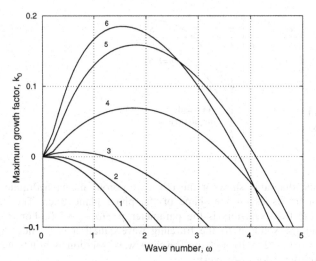

Figure 7. Plots of the critical eigenvalue as a function of wave number of distur-
bance for various values of τ_0. The curve labels 1–6 correspond to the following
values of $\tau_0 = 0$, 0.07, 0.1, 0.2, 0.5 and 1.0.

$$\frac{dA}{d\xi} + \omega^2 K(s_0)p_1 = -ks_1, \tag{41}$$

$$V\tau_0 \frac{ds_1}{d\xi} + \left(P'(s_0) + V \frac{\partial\tau(s_0, p_0)}{\partial s} \frac{ds_0}{d\xi} \right) s_1$$

$$+ \left(V \frac{\partial\tau(s_0, p_0)}{\partial p} \frac{ds_0}{d\xi} - 1 \right) p_1 = -k\tau_0 s_1 \tag{42}$$

with the same definition for A given by equation (26).

Equations (41) and (42) are in the form of a spectral problem for which
we are interested in the sign of the spectrum especially for the sign of the
critical eigenvalue k_0 if it exists. The equations are not readily amenable to
analytical solution, so a numerical solution for the eigenvalue problem will
be presented.

Results of the numerical solution of the eigenvalue problem given by
equations (41) and (42) for the critical eigenvalue $k_0(\omega)$ are presented in
Figure 7 for various values of the parameter τ_0. The other parameters
needed for the solution are the same as those used in developing Figure 2.
The result for $\tau_0 = 0$ (curve 1) is essentially the same result that would be
obtained for solving the RE eigenvalue problem given by equation (18) for
equivalent sets of porous media parameters. From Figure 7 it is observed
that the critical eigenvalue for the solution with $\tau_0 = 0$ (RE solution) is
negative for all perturbation frequencies, and therefore the RE is unconditionally stable, and this agrees with the conclusion derived in Section 3 by
analytical means. The critical eigenvalue is also negative for all frequencies

for $\tau_0 = 0.07$. However, for $\tau_0 > 0.07$ the saturation profile is sufficiently non-monotonic (see curves 3–6 in Figure 2) for the critical eigenvalues to be positive over a range of frequencies. For this set of parameters then we can establish the critical value of τ_0 for instability to be slightly less than 0.1. Also, we can say that somewhere within the range of $0.08 < \tau_0 < 0.10$ the critical eigenvalues will be negative although the saturation profiles within that range are somewhat non-monotonic. These results on the conditional stability of the RNERE have shown that the advancing flow can be unstable if the parameters fall within a specified range.

The low frequency criterion derived in Section 4.1 also applies to the results obtained here for the RNERE model. Equation (29) indicates that if the value of C is negative then the flow will be unstable. As shown earlier in this section, the pressure profiles generated by the basic solution of the RNERE model can be non-monotonic if the value of the relaxation coefficient is large enough. However, it was just shown that non-monotonicity by itself is not a sufficient condition to guarantee instability. Instead, the degree of non-monotonicity has to be large enough to lead to instability.

As a reminder we should note that the results presented up to this point have all been for the condition where the initial saturation is at or above the residual saturation. Therefore, one might argue that our results are contradictory to experimental evidence that overwhelmingly has shown that flows generally become unstable where the initial saturation is below the residual. This argument then motivates us to investigate the case where $S_{\text{init}} < S_r$. This is done in the next section.

4.2.2. RNERE Model for Initial Moisture Less than Residual

Most models of unsaturated flow are concerned with flows that occur at saturations above the residual. For the case where the initial saturation is less than the residual, an extended model is required. We will now present analyses for an extended model for unsaturated flow. Both the basic solution and a stability analysis for the basic solution will be presented.

4.2.2.1 Basic Solution for the Extended Model. The governing equations are changed slightly using saturation S as the primary variable, as opposed to effective saturation s as before. The traveling wave form of the RNERE therefore becomes (analogous to equations (30) and (31))

$$\frac{dS}{d\xi} = \frac{p - P(S)}{V\tau(S)}, \tag{43}$$

$$\frac{dp}{d\xi} = \frac{V(S - S_+) + K(S_+) - K(S)}{K(S)}. \tag{44}$$

The usual pressure-saturation and conductivity-saturation relations apply for the saturation range from residual to full saturation. To apply equations

(43) and (44) over the full range of saturation from $S=0$ to $S=1$ we need an extended model for pressure-saturation and conductivity-saturation. For this we use the approach of Rossi and Nimmo (1994) who extended the modified Brooks–Corey model for the range $0 \leqslant S \leqslant S_j$, except that we use the conventional Brooks–Corey function for the range $S_j \leqslant S \leqslant 1$. The parameter S_j is the value of saturation joining the two ranges of saturation. The pressure-saturation model now becomes

$$P(S) = -p_0 \exp(-aS), \quad 0 \leqslant S \leqslant S_j,$$
$$P(S) = -\left(\frac{S - S_r}{1 - S_r}\right)^{-\beta}, \quad S_j \leqslant S \leqslant 1. \tag{45}$$

The conductivity-saturation relation is given by

$$K(S) = \sqrt{S}[I(S)/I(1)]^2, \tag{46}$$

where

$$I(S) = 1 - \frac{1}{ap_0} e^{aS}, \quad 0 < S < S_j, \tag{47}$$

$$I(S) = 1 - \frac{1}{ap_0} e^{aS_j} + \frac{1 - S_r}{\beta + 1}\left[\left(\frac{S_j - S_r}{1 - S_r}\right)^{\beta+1} - \left(\frac{S - S_r}{1 - S_r}\right)^{\beta+1}\right], \quad S_j < S \leqslant 1. \tag{48}$$

The computed parameter a and S_j provide first-order continuity for these functions. The S_j is set at a saturation slightly above S_r.

For the relaxation coefficient we used relation (34) with $\delta = 0$. Asymptotic analysis showed that for bounded solutions as $S_{\text{init}} \to 0$ we need $\gamma > 0.5$.

Parameter values for the model were chosen to fit the experimental data of Bauters et al. (2000), who performed experiments to evaluate the effect of initial saturation on flow stability. They examined initial water contents ranging from air-dry conditions to residual water content. For the analysis to follow we examined a subset of their experiments, using initial water contents θ_{init} :0.001, 0.01, 0.02, 0.03, 0.04 and 0.047. The residual water content for their porous media was 0.047 and saturated water content was 0.348. Parameter values obtained from the published moisture retention data were $S_r = 0.135$ and $\beta = 0.18$, while for the conductivity function we used $p_0 = 10^5$ m and $a = 1.5$. We also calibrated the relaxation coefficient relation using the published water content and pressure profiles for the experimental run for $\theta_{\text{init}} = 0.001$, and obtained $\tau_0 = 1.5$ and $\gamma = 1.5$.

The saturation profiles resulting from the solution to equations (43) and (44) with the specified parameters are shown in Figure 8 for the various values of S_{init} corresponding to the various initial water contents. The profiles for the very dry initial conditions are clearly non-monotonic, while

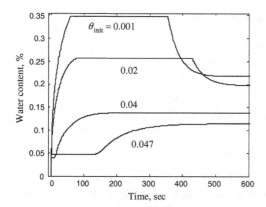

Figure 8. Water content as a function of dimensional time at any particular point within the flow domain. The cases $\theta_{init} = 0.01$ and 0.03 represented in Table I are not shown.

the profiles for the higher initial saturations are essentially monotonic. The characteristics of these profiles with respect to flow stability will be discussed in the next section.

4.2.2.2. Stability Analysis for the Extended Model. The analysis of flow stability for the extended RNERE model follows the same procedures as for the RNERE model outlined in Section 4.2.1. The resulting spectral problem is essentially the same as that shown by equations (41) and (42). The main difference between the current model and the previous model is in the parameters of the system due to the extension to dry conditions. As before, the complexity of the resulting perturbation equation makes it necessary to solve the eigenvalue problem by numerical means.

When we attempted the numerical evaluation of the spectral problem for the extended RNERE we found the numerical procedures used previously to be inadequate to get accurate results. The reason for the numerical difficulty is the extremely steep wetting front that develops for the basic solution in the case of the dry initial condition (see Figure 8). Therefore, to derive some useful results the direct evaluation of the spectral problem was abandoned for the time being and the low-frequency criterion analysis derived by Egorov *et al.* (2003) and outlined in Section 4.1 was applied. Revisiting the direct spectral problem will have to await further investigation into more accurate means to solve the steep front problem.

The low-frequency criterion for flow stability was given by equations (28) and (29). These equations state that when the pressure profile is sufficiently non-monotonic the value of C will be negative, and this will then lead to a positive value of k_0, indicating the flow will be unstable. So the determination of the stability of flows generated from dry initial conditions is to simply evaluate expression (28) with the $K(s_0)\mathrm{d}p_0/\mathrm{d}\xi$ derived from the

Table I. Finger widths as a function of initial water content observed in the experiments of Bauters *et al.* (2000), and corresponding values of *C* evaluated by Equation (28)

θ_{init}	0.001	0.01	0.02	0.03	0.04	0.047
Finger (cm)	2.5	1.25	3.0	7.5	11.0	30.0
C	−2.46	−2.85	−0.96	−0.087	−0.093	0.009

traveling wave solution for a specific set of porous media parameters, and if the value of *C* is negative to then conclude that the flow will be unstable.

The value of *C* evaluated with equation (28) for the various initial water contents is presented in Table I. Finger widths reported by Bauters *et al.* (2000) for the corresponding cases are also listed. The experimental chamber had a width of 30 cm. It is observed from the table that as the initial water content approached the residual of 0.047 the finger width increased to the size of the experimental chamber.

From the values presented in the table it appears that the only case of stable flow ($C > 0$) occurs for $\theta_{init} = 0.047$. All the other cases have negative values of *C*, indicating instability of flow. The finger widths observed by Bauters *et al.* (2000) are in agreement with this result. The largest finger width was observed for $\theta_{init} = 0.047$, with the finger width being at least equal to the width of the chamber. The next largest finger width was 11.0 cm for the case with $\theta_{init} = 0.04$. While the plot in Figure 8 for $\theta_{init} = 0.04$ does not appear to be non-monotonic, it was sufficiently non-monotonic so that the value of *C* was negative in the evaluation of equation (28). For the other two cases shown in Figure 8, $\theta_{init} = 0.001$ and $\theta_{init} = 0.02$, the saturation profiles are clearly non-monotonic, and this is manifested in the values of *C* presented in Table I.

5. Summary and Conclusion

We have presented several alternative forms of the equations for flow in unsaturated porous media. All of the various forms contain two coupled equations, the mass balance equation derived from a combination of conservation of mass and a linear flux law, and a relationship for the pressure that appears in the mass balance equation. The differences between the different forms of equations are contained completely in the definition of the pressure function. For the conventional equation for unsaturated flow, given by the RE, the pressure function is given by the equilibrium capillary pressure – saturation relationship. A more general equation system is derived using a generalized non-equilibrium relationship of the pressure function, and this is called the NERE. A specific form of the non-equilibrium relationship is given by a relaxation function expressed by a first-order rate process. The equation system using this relaxation function is referred to as the RNERE.

The traveling wave solution was studied for each of these models and stability analysis of these traveling wave solutions was performed. The characteristics of the traveling wave solutions for the RE and the RNE-RE were analyzed in detail, and expressed in terms of the monotonicity or non-monotonicity of the traveling wave saturation and pressure profiles. These solutions were derived for the standard case where the initial saturation is above residual saturation. The RNERE was also extended to the case where the initial saturation is less than the residual, since this corresponds better with conventional laboratory experiments on unstable flows.

For the linear stability analysis the solution to the system of equations in the traveling wave variable yielded a basic solution that was then perturbed by infinitesimal fluctuations. Stability of the perturbed flow was then analyzed by the method of spectral analysis. The stability of the NERE was analyzed at low-frequency only because of the general form of the non-equilibrium pressure function, and the resulting criterion is called the LFC. The stability analysis for the RNERE for initially dry conditions was also limited to the LFC because of the numerical difficulty to accurately solve the perturbation equation with extremely sharp fronts.

From the analyses presented here we can conclude the following.

(1) The RE is unconditionally stable to any perturbation, whether infinitesimal or finite in magnitude, in homogeneous or heterogeneous porous media. The traveling wave saturation and pressure profiles for the RE are monotonic for standard type boundary conditions.

(2) The analysis of the stability of the NERE model using the LFC shows that infiltrating flows governed by the NERE can become unstable for conditions where the flow profile is sufficiently non-monotonic.

(3) The saturation and pressure profiles for the RNERE model were found to be non-monotonic for a sufficient large value of the relaxation coefficient parameter τ_0. The larger this parameter the larger is the degree of non-monotonicity of the profiles. Assessments of stability over a wide range of perturbation frequencies were completed showing that flows are stable even for slight non-monotonicity, but transition to unstable as the degree of non-monotonicity increases.

(4) For initially dry conditions, the LFC assessment of the RNERE indicated unstable flow for initial conditions from absolutely dry up to near residual saturation. With initial saturation equal to the residual, the LFC indicated that the flow may be stable. These stability assessments were in good agreement with the stability experiments of Bauters et al. (2000).

Acknowledgements

The authors wish to acknowledge support under NATO Collaborative Linkage Grant 978242 and Minnesota Agricultural Experiment Station Project 12-044. Partial support for Andrey Egorov and Rafail Dautov was received from the Russian Foundation for Basic Research Grant 03-01-96237. The authors also gratefully acknowledge helpful comments offered by reviewers of the original manuscript.

References

Baker, R. S. and Hillel, D.: 1990, Laboratory tests of a theory of fingering during infiltration into layered soils, *Soil Sci. Soc. Am. J.* **54**, 20–30.

Bauters, T. W. J., DiCarlo, C. A., Steenhuis, T. S. and Parlange, J.-Y.: 2000, Soil water content dependent wetting front characteristics in sands, *J. Hydrol.* 231–232, 244–254.

Brooks, R. H. and Corey, A. T.: 1964, Hydraulic properties of porous media, Hydrology Paper No. 3, Colorado State University, Fort Collins

Carmona, R. A. and Lacroix, J.: 1999, *Spectral Theory of Random Schrodinger Operators*, Birkhauser, Boston.

Chuoke, R. L., van Meurs, P. and van der Poel, C.: 1959, The instability of slow immiscible, viscous liquid-liquid displacements in porous media, *Trans. Am. Inst. Min. Metall. Pet. Eng.* **216**, 188–194.

Cuesta, C., van Duijn, C. J. and Hulshof, J.: 2000, Infiltration in porous media with dynamic capillary pressure: traveling waves, *Eur. J. Appl. Math.* **11**, 381–397.

Dahle, H. K., Celia, M. A., Hassanizadeh, S. M. and Karlsen, K. H.: 2002, A total pressure-saturation formulation of two-phase flow incorporating dynamic effects in the capillary-pressure-saturation relationship, in: S. M. Hassanizadeh, R. J. Schotting, W. G. Gray and G. F. Pinder (eds.), *Computational Methods in Water Resources*, Vol. 2, No. 47 in series of Developments in Water Science, Elsevier, Amsterdam, pp. 1067–1074.

Deinert, M., Parlange, J.-Y., Steenhuis, T. S., Unlu, K., Selker, J. and Cady, K. B.: 2002, Real-time measurement of water profiles in a sand using neutron radiograph, in: *Proceedings of the 22nd Ann. Am. Geophs. Union Hydrol. Days*, pp. 56–63.

De Rooij, G. H. and Cho, H.: 1999, Modelling solute leaching during fingered flow by integrating and expanding various theoretical and empirical concepts, *Hydrol. Sci. J.* **44**, 447–465.

Diment, G. A. and Watson, K. K.: 1983, Stability analysis of water movement in unsaturated porous materials, 2. Numerical studies, *Water Resour. Res.* **19**, 1002–1010.

Diment, G. A. and Watson, K. K.: 1985, Stability analysis of water movement in unsaturated porous materials, 3. Experimental studies, *Water Resour. Res.* **21**, 979–984.

Diment, G. A., Watson, K. K. and Blennerhassett, P. J.: 1982, Stability analysis of water movement in unsaturated porous materials, 1. Theoretical considerations, *Water Resour. Res.* **18**, 1248–1254.

Du, X., Yao, T., Stone, W. D. and Hendrickx, J. M. H.: 2001, Stability analysis of the unsaturated flow equation. 1. Mathematical derivation, *Water Resour. Res.* **37**, 1869–1874.

Egorov, A. G., Dautov, R. Z., Nieber, J. L. and Sheshukov, A. Y.: 2002, Stability analysis of traveling wave solution for gravity-driven flow, in: S. M. Hassanizadeh, R. J. Schotting, W. G. Gray and G. F. Pinder (eds.), *Computational Methods in Water Resources*, Vol. **1**, No. 47 in series of Developments in Water Science, Elsevier, Amsterdam, pp. 121–128.

Egorov, A. G., Dautov, R. Z., Nieber, J. L. and Sheshukov, A. Y.: 2003, Stability analysis of gravity-driven infiltrating flow, *Water Resour. Res.* **39**, 1266, doi:10.1029/2002WR001886.

Glass, R. J., Parlange, J.-Y. and Steenhuis, T. S.: 1988, Wetting front instability as a rapid and far reaching hydrologic process in the vadose zone, in: Germann, P. F. (ed.), *Rapid and Far Reaching Hydrologic Processes in the Vadose Zone, J. Conam. Hydrol.* **3**, 207–226.

Glass, R. J., Steenhuis, T. S and Parlange, J.-Y.: 1989a, Mechanism for finger persistence in homogeneous, unsaturated porous media: theory and verification, *Soil Sci.* **148**, 60–70.

Glass, R. J., Steenhuis, T. S and Parlange, J.-Y.: 1989b, Wetting front instability, 1. Theoretical discussion and dimensional analysis, *Water Resour. Res.* **25**, 1183–1194.

Glass, R. J., Steenhuis, T. S and Parlange, J.-Y.: 1989c, Wetting front instability, 2. Experimental determination of relationships between system parameters and two-dimensional unstable flow field behavior in initially dry porous media, *Water Resour. Res.* **25**, 1195–1207.

Glass, R. J., Parlange, J.-Y. and Steenhuis, T. S.: 1991, Immiscible displacement in porous media: stability analysis of three-dimensional, axisymmetric disturbances with application to gravity-driven wetting front stability, *Water Resour. Res.* **27**, 1947–1956.

Hassanizadeh, S. M., Celia, M. A. and Dahle, H. K.: 2002, Dynamic effect in the capillary pressure-saturation relationship and its impacts on unsaturated flow, *Vadose Zone J.* **1**, 28–57.

Hassanizadeh, S. M. and Gray, W. G.: 1993, Thermodynamic basis of capillary pressure in porous media, *Water Resour. Res.* **29**, 3389–3405.

Hill, D. E. and Parlange, J.-Y.: 1972, Wetting front instability in layered soils, *Soil Sci. Soc. Am. Proc.* **36**, 697–702.

Kapoor, V.: 1996, Criterion for instability of steady-state unsaturated flows, *Transp. Porous Media* **25**, 313–334.

Kirkham, D. and Feng, L.: 1949, Some tests of the diffusion theory, and laws of capillary flow in soils, *Soil Sci.* **67**, 29–40.

Liu, Y., Steenhuis, T. S and Parlange, J.-Y.: 1994a, Closed-form solution for finger width in sandy soils at different water contents, *Water Resour. Res.* **30**, 949–952.

Liu, Y., Steenhuis, T. S and Parlange, J.-Y.: 1994b, Formation and persistence of fingered flow fields in coarse grained soils under different moisture contents, *J. Hydrol.* **159**, 187–195.

Mualem, Y.: 1974, A conceptual model of hysteresis, *Water Resour. Res.* **10**, 514–520.

Nieber, J. L.: 2001, The relation of preferential flow to water quality, and its theoretical and experimental quantification, In: D. Bosch and K. W. King (eds), *Preferential Flow, Water Movement and Chemical Transport in the Environment,* American Society of Agricultural Engineers, St. Joseph, MI. pp. 1–10.

Nielson, D. R., Biggar, G. and Davidson, G.: 1962, Experimental consideration of diffusion analysis in unsaturated flow problems, *Soil Sci. Soc. Am. Proc.* **26**, 107–111.

Otto, F.: 1996, L^1-contraction and uniqueness for quasilinear elliptic-parabolic equations, *J. Diff. Eq.* **131**, 20–38.

Otto, F.: 1997, L^1-contraction and uniqueness for unstationary saturated-unsaturated water flow in porous media, *Adv. Math. Sci. Appl.* **7**, 537–553

Panfilov, M.: 1998, Upscaling two-phase flow in double porosity media: Nonuniform homogenization, in: J. M. Crolet and M. E. Hatri (eds), *Recent Advances in Problems of Flow and Transport in Porous Media,* pp. 195–215.

Parlange, J.-Y. and Hill, D. E.: 1976, Theoretical analysis of wetting front instability in soils, *Soil Sci.* **122**, 236–239.

Philip, J. R.: 1957, The theory of infiltration, 4. Sorptivity analysis of algebraic infiltration equations, *Soil Sci.* **84**, 257–264.

Philip, J. R.: 1975a, Stability analysis of infiltration, *Soil Sci. Soc. Am. Proc.* **39**, 1042–1049.
Philip, J. R.: 1975b, The growth of disturbances in unstable infiltration flows, *Soil Sci. Soc. Am. Proc.* **39**, 1049–1053.
Raats, P. A. C.: 1973, Unstable wetting fronts in uniform and nonuniform soils, *Soil Sci. Soc. Am. Proc.* **37**, 681–685.
Rawlins, S. L. and Gardner, W. H.: 1963, A test of the validity of the diffusion equation for unsaturated flow of soil water, *Soil Sci. Soc. Am. Proc.* **27**, 507–511.
Rossi, C. and Nimmo, J. R.: 1994, Modeling of soil water retention from saturation to oven dryness, *Water Resour. Res.* **30**, 701–708.
Saffman, P. G. and Taylor, G.: 1958, The penetration of a fluid into a porous media or Hele-Shaw cell containing a more viscous fluid, *Proc. R. Soc. London, Ser. A*, **245**, 312–329.
Selker. J. S., Steenhuis, T. S. and Parlange, J.-Y.: 1992a, Wetting front instability in homogeneous sandy soils under continuous infiltration, *Soil Sci. Soc. Am. J.* **56**, 1346–1350.
Selker, J. S., Parlange, J.-Y. and Steenhuis, T. S.: 1992b, Fingered flow in two dimensions, 1, Measurement of matric potential, *Water Resour. Res.* **28**, 2513–2521.
Sililo, O. T. N. and Tellam, J. H.: 2000, Fingering in unsaturated flow: a qualitative review with laboratory experiments on heterogeneous systems, *Ground Water* **38**, 864–871.
Smiles, D. E., Vachaud, G. and Vauclin, M.: 1971, A test of the uniqueness of the soil moisture characteristic during transient, non-hysteretic flow of water in rigid soil, *Soil Sci. Soc. Am. Proc.* **35**, 535–539.
Smith, W. O.: 1967, Infiltration in sand and its relation to ground water recharge, *Water Resour. Res.* **3**, 539–555.
Starr, J. L., DeRoo, H. C., Frink, C. R., and Parlange, J. Y.: 1978, Leaching characteristics of a layered field soil, *Soil Sci. Soc. Am. J.* **42**, 386–391.
Tabuchi, T.: 1961, Infiltration and ensuing percolation in columns of layered glass particles packed in laboratory, *Nogyo Dobuku kenkyn, Bessatsu (Trans. Agr. Eng. Soc. Japan)*, **1**, 13–19. (in Japanese with summary in English).
Topp G. C., Klute, A. and Peters, D. B.: 1967, Comparison of water content-pressure head data obtained by equilibrium, steady-state, and unsteady-state methods, *Soil Sci. Soc. Am. Proc.* **31**, 312–314.
Ursino, N.:2000, Linear stability analysis of infiltration, analytical and numerical solution, *Transp. Porous Media* **38**, 261–271.
van Genuchten, M. T.: 1980, A closed form equation for predicting the hydraulic conductivity of unsaturated soil, *Soil Sci. Soc. Am. J.* **44**, 892–898.
Wang, Z., Feyen, J. and Elrick, D. E.: 1998a, Prediction of fingering in porous media, *Water Resour. Res.* **34**, 2183–2190.
Wang, Z., Feyen, J., van Genuchten, M. Th. and Nielsen D. R.: 1998b, Air entrapment effects on infiltration rate and flow instability, *Water Resour. Res.* **34**, 213–222.
Wildenschild, D., Hopmans, J. W. and Simunek, J.: 2001, Flow rate dependence of soil hydraulic characteristics, *Soil Sci. Soc. Am. J.* **65**, 35–48.

Transp Porous Med (2005) 58:173–189
DOI 10.1007/s11242-004-5474-4

Analytic Analysis for Oil Recovery During Counter-Current Imbibition in Strongly Water-Wet Systems

ZOHREH TAVASSOLI, ROBERT W. ZIMMERMAN and
MARTIN J. BLUNT*
Department of Earth Science and Engineering, Imperial College, London, UK

(Received: 23 October 2003; in final form: 24 March 2004)

Abstract. We study counter-current imbibition, where a strongly wetting phase (water) displaces non-wetting phase spontaneously under the influence of capillary forces such that the non-wetting phase moves in the opposite direction to the water. We use an approximate analytical approach to derive an expression for saturation profile when the viscosity of the non-wetting phase is non-negligible. This makes the approach applicable to water flooding in hydrocarbon reservoirs, or the displacement of non-aqueous phase liquid (NAPL) by water. We find the recovery of non-wetting phase as a function of time for one-dimensional flow. We compare our predictions with experimental results in the literature. Our formulation reproduces experimental data accurately and is superior to previously proposed empirical models.

Key words: imbibition, counter-current flow, fractured reservoirs

Nomenclature

a, b, c, f	exponents.
A	area, L^2, m^2.
$A(t), B(t)$	functions of time, dimensionless.
C	integration constant, dimensionless.
J	Leverett J-function (dimensionless capillary pressure).
J'	J-function gradient at $S = 1$.
J^*	Prefactor in Equation (51).
K	permeability, L^2, m^2 or D.
K_r	relative permeability.
L, l	length, L, m.
m, n	exponents in van Genutchen and Lenhard & Parker models.
P_c	capillary pressure, $mL^{-1} t^{-2}$, Pa.
P_c^*	minimum capillary pressure, $mL^{-1} t^{-2}$, Pa.

---

*Author for correspondence e-mail: m.blunt@imperial.ac.uk

R	recovery.
S	saturation.
t	time, T, s.
V	volume, L^3, m^3.
$\alpha, \beta, \eta, \kappa, \gamma, \eta$	dimensionless rate constants.
χ	rate constant, T^{-1}, s^{-1}.
ϕ	porosity, dimensionless.
λ	mobility, M^{-1} Lt, $Pa^{-1} \cdot s^{-1}$.
μ	viscosity, $ML^{-1} t^{-1}$, $Pa \cdot s$.
ρ	density, ML^{-3}, $kg \cdot m^{-3}$.
σ	interfacial tension, $L^{-1} t^{-2}$, Nm^{-1}.
AFO	All-Faces-Open boundary condition.
OEO	One-End-Open boundary condition.
TEO	Two-End-Open boundary condition.
TEC	Two-End-Closed boundary condition.

Subscripts

D	dimensionless.
i	initial.
o	oil.
r	residual.
t	total.
w	water.
∞	ultimate (at infinite time).

Superscripts

max	maximum.

1. Introduction

One of the most important recovery mechanisms in hydrocarbon reservoirs is spontaneous imbibition, where capillary pressure causes water to invade into water-wet and mixed-wet rock containing oil. In fractured reservoirs, injected water flows rapidly through the fracture network and then can imbibe into low permeability matrix that contains the majority of the oil. In many cases this is a counter-current imbibition process: water and oil flow in opposite directions with the sum of the water and oil Darcy velocities being zero. The same situation is seen in fractured aquifers polluted by non-aqueous phase liquids (NAPLs) – again the imbibition of water can recover NAPL present in the less permeable matrix.

It is possible to study counter-current imbibition at the core scale in the laboratory. A core sample containing oil is surrounded by water and the recovery of oil is measured as a function of time. In this set-up the only force mediating the displacement is capillary pressure with water and oil flowing in opposite directions. For a recent review of these experiments see Morrow and Mason (2001). The recovery of oil can be fitted by a simple exponential function of time (Aronofsky et al., 1958). Zhang et al.

(1996) proposed the following expression that matched a range of imbibition experiments on samples with different geometry and fluid properties:

$$R = R_\infty \left(1 - e^{-\alpha t_{DE}}\right) \tag{1}$$

where R is the recovery and R_∞ is the ultimate recovery. The subscript E in the dimensionless time t_{DE} stands for empirical. The constant α is approximately 0.05 and the dimensionless time is defined by:

$$t_{DE} = t \sqrt{\frac{K}{\phi}} \frac{\sigma}{\sqrt{\mu_w \mu_o}} \frac{1}{L_c^2} \tag{2}$$

where σ is the interfacial tension and L_c is a characteristic or effective length given by:

$$L_c = \frac{V}{\sum\limits_{i=1}^{n} \frac{A_i}{l_i}} \tag{3}$$

where V is the matrix block volume, A_i is the area open to flow in the ith direction and l_i is the distance from the open surface to a no-flow boundary. Equation (1) can be re-written in terms of the average water saturation in the core, $\bar{S}_w$:

$$\frac{R}{R_\infty} = \frac{\bar{S}_w - S_{wi}}{1 - S_{or} - S_{wi}} \tag{4}$$

where S_{wi} is the initial water saturation (typically a connate or irreducible value) and S_{or} is the residual oil saturation, or, more precisely, the oil saturation that is reached after spontaneous imbibition only. Defining a normalized saturation:

$$S = \frac{S_w - S_{wi}}{1 - S_{wi} - S_{or}}; \quad 1 \geqslant S \geqslant 0. \tag{5}$$

$$\bar{S} = 1 - e^{-\alpha t_{DE}} \tag{6}$$

For mixed-wet systems, an approximately exponential recovery is also seen, but with values of α that are 100–10,000 times lower than for strongly water-wet media (Morrow and Mason, 2001).

   Zimmerman and Bodvarsson (1989, 1991) used an approximate analytical technique, the integral method, to study counter-current imbibition in one dimension. Later Zimmerman et al. (1990) extended the approach to study imbibition into blocks of various shapes and sizes and showed good comparisons with numerical solutions. Zimmerman and Bodvarsson (1989, 1991), Zimmerman et al. (1990) considered water/air systems with the assumption that the displaced phase (air) had a negligible viscosity. For hydrocarbon or NAPL problems, however, the non-wetting phase viscosity cannot be neglected.

Barenblatt *et al.* (1990) derived expressions for the early and late-time recovery from counter-current imbibition where both phases have non-negligible viscosity. At late times, using a similar technique to Zimmerman and Bodvarsson (1989, 1991), they found an exponential variation of recovery with time, Equation (1), but with a different dimensionless time.

Many other authors have also studied this problem. Pooladi-Darvish and Firoozabadi (2000) constructed numerical solutions for one-dimensional counter-current flow using power-law forms of the relative permeability. They demonstrated that the saturation gradient at the inlet is infinite, but did not compare the results to experiment or propose a closed-form expression for the average saturation as a function of time. Kashchiev and Firoozabadi (2003) found analytic solutions for early time imbibition (before water reaches any boundaries) in different flow geometries and compared their results with numerical solutions and experiment. Reis and Cil (1993) found a closed-form analytical model for capillary imbibition in one dimension. They assumed that the oil and water saturations behind the imbibition front varied linearly with distance, the capillary pressure was a linear function of water saturation, and the ratio of the capillary pressure to the sum of the oil and water mobilities was constant. Chen *et al.* (1995) derived integral solutions for late times in one and two dimensions by extending the approach used in Chen *et al.* (1990) for the infinite-acting flow period. Using a more empirical approach, Zhou *et al.* (2002) suggested a new scaling group, similar to that derived by Barenblatt *et al.* (1990), that unlike Equation (2) tended to the correct limit for a negligible non-wetting phase viscosity. They validated their expression through comparison with experiment on diatomite cores. A new approach to this problem was proposed by Barenblatt *et al.* (2003) who used an extended theory of multiphase flow with a relaxation time to explain counter-current imbibition. In our work, we will assume that a traditional Darcy-like formulation of the flow equations is sufficient for the analysis.

In this paper we will follow the approach of Zimmerman and Bodvarsson (1989, 1991) and Zimmerman *et al.* (1990) to analyze counter-current imbibition in one dimension for oil/water or NAPL/water systems. We propose both early and late time solutions and make fewer approximations about the functional form of the saturation profile, capillary pressures and relative permeabilities than previous authors. We will derive a simple analytical form for the average water saturation that gives a better match to experimental data than the exponential expression, Equation (6). The key difference between this work and that of Barenblatt *et al.* (1990) is the treatment of the boundary conditions – for strongly water-wet systems the water mobility is zero at the inlet. As we show later, this leads to an infinite saturation gradient and results in a different functional form from the exponential recovery previously proposed.

2. Analysis

Conservation of water volume in one dimension with no overall flow can be expressed as follows (Barenblatt et al., 1990):

$$\phi \frac{\partial S_w}{\partial t} + \frac{\partial}{\partial x}\left(\frac{\lambda_w \lambda_o}{\lambda_t} K \frac{\partial P_c}{\partial S_w}\frac{\partial S_w}{\partial x}\right) = 0 \tag{7}$$

where K is the permeability, P_c is the capillary pressure, the mobility $\lambda = K_r/\mu$ where K_r is the relative permeability and $\lambda_t = \lambda_w + \lambda_o$. In what follows, for simplicity, we will refer to the non-wetting phase as 'oil' represented by the subscript o. The other terms are defined in the nomenclature.

We rewrite Equation (7) in terms of dimensionless variables: the normalized saturation, S, Equation (5) and a dimensionless length $x_D = x/L$. The boundary conditions for flow in $1 \geqslant x_D \geqslant 0$ are: at $x_D = 0$, $S = 1$; and at $x_D = 1$ there is no flux, meaning that $\partial S/\partial x_D = 0$. The conservation equation, assuming a constant permeability, is then:

$$\frac{\partial S}{\partial t} + \frac{K}{\phi(1 - S_{wi} - S_{or})L^2}\frac{\partial}{\partial x_D}\left(\frac{\lambda_w \lambda_o}{\lambda_t}\frac{\partial P_c}{\partial S}\frac{\partial S}{\partial x_D}\right) = 0 \tag{8}$$

Rather than attempt a solution of the non-linear Equation (8) directly, we will construct a solution of the weak or integral form of the equation (Zimmerman and Bodvarsson, 1989, 1991; Zimmerman et al., 1990):

$$\int_0^1 \left[\frac{\partial S}{\partial t} + \frac{K}{\phi(1 - S_{wi} - S_{or})L^2}\frac{\partial}{\partial x_D}\left(\frac{\lambda_w \lambda_o}{\lambda_t}\frac{\partial P_c}{\partial S}\frac{\partial S}{\partial x_D}\right)\right]dx_D = 0 \tag{9}$$

which gives:

$$\frac{\partial \bar{S}}{\partial t} = \frac{K}{\phi(1 - S_{wi} - S_{or})L^2}\left(\frac{\lambda_w \lambda_o}{\lambda_t}\frac{\partial P_c}{\partial S}\frac{\partial S}{\partial x_D}\right)\Bigg|_{x_D=0} \tag{10}$$

where:

$$\bar{S} = \int_0^1 S\,dx_D \tag{11}$$

The solution to Equation (10) is controlled by the functional form of the capillary pressure and mobilities at the inlet, where S tends to 1. For convenience we will assume the following Corey-type functional forms for the mobilities and the imbibition capillary pressure (Bear, 1972; Dullien, 1992):

$$\lambda_w = \lambda_w^{max} S^a \tag{12}$$

$$\lambda_o = \lambda_o^{max} (1 - S)^b \tag{13}$$

$$P_c = \sigma \sqrt{\frac{\phi}{K}} J(S_w) \tag{14}$$

$$\left.\frac{\partial P_c}{\partial S_w}\right|_{S_w=1-S_{or}} = -\sigma \sqrt{\frac{\phi}{K}} J' = \frac{1}{(1 - S_{wi} - S_{or})} \left.\frac{\partial P_c}{\partial S}\right|_{S=1} \tag{15}$$

where $\lambda_w^{max} = k_{rw}^{max}/\mu_w$ and $\lambda_o^{max} = k_{ro}^{max}/\mu_o \cdot a, b > 0$. In Equation (14) we have assumed Leverett J-function scaling (Bear, 1972; Dullien, 1992) where $J(S_w)$ is a dimensionless capillary pressure and $-J'$ is the dimensionless gradient of the capillary pressure at $S = 1$.

We now rewrite Equation (10) with $1 - S = \varepsilon$ and take the limit as $\varepsilon \to 0, x_D \to 0$:

$$\frac{\partial \bar{S}}{\partial t} = -\sqrt{\frac{K}{\phi} \frac{k_{ro}^{max} \sigma J'}{\mu_o L^2}} \varepsilon^b \left.\frac{\partial S}{\partial x_D}\right|_{x_D=0} \tag{16}$$

Now define a dimensionless time t_D:

$$t_D = t \sqrt{\frac{K}{\phi} \frac{\sigma}{\mu_o L^2}} \tag{17}$$

and Equation (16) becomes:

$$\frac{\partial \bar{S}}{\partial t_D} = k_{ro}^{max} J' \varepsilon^b \left.\frac{\partial S}{\partial x_D}\right|_{x_D=0} \tag{18}$$

Since the oil mobility vanishes for $\varepsilon = 0$, it is necessary that either $\partial S/\partial x_D$ or $\partial P_c/\partial S$ diverges at $x_D = 0$ to obtain sensible solutions to Equation (18). If we assume that the capillary pressure gradient and therefore J' is finite, this means that the saturation gradient must be infinite at $x_D = 0$, in accordance with previous numerical solutions of the problem (Pooladi-Darvish and Firoozabadi, 2000). Physically the inlet boundary condition states that the oil saturation is at its residual value where its mobility is zero. However, oil must be able to flow out of the system. The only way to resolve this impasse is to allow an infinite saturation gradient such that the oil flow rate is finite.

Many empirical expressions for capillary pressure, particularly those widely used in the hydrology literature due to van Genutchen and Lenhard and Parker, have an infinite saturation gradient at $S = 1$ (van Genutchen, 1980; Lehnard and Parker, 1987a,b,c; Parker et al., 1987). The theory for these cases follows a very similar development to that given below and is outlined in the Appendix.

The time-scale for imbibition, given by t_D, Equation (17), is similar to Equation (2), except for the scaling with viscosity. In our analysis the imbibition time is inversely proportional to the oil viscosity and is independent of the water viscosity. Zhang *et al.* (1996) suggested that an imbibition time inversely proportional to the geometric average of the oil and water viscosities more accurately matched experiment. We will test this against experimental data later in the paper.

2.1. EARLY TIME SOLUTION

First we propose an early time solution, before the advancing water front reaches the boundary. This solution is also equivalent to imbibition in a semi-infinite medium. We write down a simple analytical form for the spatial variation of the saturation and then find the time-dependent coefficients that obey Equation (16) and the boundary conditions.

$$
\begin{aligned}
S(x_D, t_D) &= 1 - A(t_D)x_D^f; \quad x_D \leqslant x_D^0, \ 0 < f < 1 \\
&= 0; \qquad\qquad\quad x_D \geqslant x_D^0
\end{aligned}
\tag{19}
$$

This automatically obeys the boundary condition at $x_D = 0$. $x_D^0(t_D)$ is the length at which the saturation first becomes zero. Since $S\left(x_D^0, t_D\right) = 0$, from Equation (19) we find:

$$
x_D^0 = A(t_D)^{-1/f}
\tag{20}
$$

and the average saturation is:

$$
\bar{S}(t_D) = \frac{f}{1+f}x_D^0 = \frac{f}{(1+f)A(t_D)^{1/f}}
\tag{21}
$$

We now use Equation (18) to find $A(t_D)$. From Equation (19):

$$
\varepsilon = A(t_D)x_D^f
\tag{22}
$$

Hence in Equation (18):

$$
\frac{A(t_D)^{-1/f-1}}{1+f}\frac{\partial A(t_D)}{\partial t_D} = -k_{ro}^{\max} J' f \varepsilon^{b+1-1/f} A(t_D)^{1/f}
\tag{23}
$$

For the right-hand-side of Equation (23) to be finite in the limit $\varepsilon \to 0$, we require:

$$
f = \frac{1}{1+b}
\tag{24}
$$

This specifies f and gives acceptable solutions for any $b > 0$. Then Equation (23) becomes:

$$\frac{A(t_D)^{-1-2/f}}{f(1+f)}\frac{\partial A(t_D)}{\partial t_D} = -k_{ro}^{max}J' \tag{25}$$

Integrating once:

$$A(t_D) = \frac{1}{(\beta t_D + C)^{f/2}} \tag{26}$$

where:

$$\beta = \frac{2(2+b)}{1+b}k_{ro}^{max}J' \tag{27}$$

and C is a constant.

Then from Equation (21) and choosing the constant $C = 0$ to obey the initial condition $x_D^0(0) = \bar{S}(0) = 0$:

$$\bar{S}(t_D) = (\gamma t_D)^{1/2} \tag{28}$$

where:

$$\gamma = \frac{\beta}{(2+b)^2} = \frac{2}{(2+b)(1+b)}k_{ro}^{max}J' \tag{29}$$

The leading edge of the water front advances as (Equations (20) and (26)):

$$x_D^0(t_D) = (\beta t_D)^{1/2} \tag{30}$$

This solution is valid for $t_D \le t_{D1}$ until the water reaches the boundary, when $x_D^0(t_D = t_{D1}) = 1$, where:

$$t_{D1} = \frac{1}{\beta} \tag{31}$$

Notice that the average saturation and the location of the leading edge of the water front increases as the square root of time. This is readily explained, since the advance rate is proportional to the capillary pressure gradient which is inversely proportional to the distance the water front has already traveled. This scaling of average saturation has been noted by other authors (Zimmerman and Bodvarsson, 1989, 1991; Barenblatt et al., 1990; McWhorter and Sunada, 1990; Kashchiev and Firoozabadi, 2003) and confirmed experimentally (Rangel-German and Kovscek, 2002; Zhou et al., 2002). However, the dimensionless rate constant γ or β is different from that derived in other work – in particular the behavior is controlled by the inlet boundary condition in this analysis.

2.2. LATE TIME SOLUTION

At late times, $t_D \geqslant t_{D1}$, we propose a similar functional form for the saturation profile:

$$S(x_D, t_D) = 1 - A(t_D)x_D^f + B(t_D)x_D; \quad 1x_D1, \ 0 < f < 1 \tag{32}$$

The extra linear term in x_D is added to fulfill the no flux constraint at $x_D = 1$, $\partial S/\partial x_D|_{x_D=1} = 0$. Hence:

$$B(t_D) = f A(t_D) \tag{33}$$

and:

$$S(x_D, t_D) = 1 - A(t_D)\left(x_D^f - f x_D\right) \tag{34}$$

The average saturation is:

$$\bar{S}(t_D) = 1 - A(t_D)\left(\frac{1}{f+1} - \frac{f}{2}\right) \tag{35}$$

We now use Equation (18) to find $A(t_D)$ in the limit as $\varepsilon \to 0, x_D \to 0$ as before and using Equation (22):

$$\left(\frac{1}{f+1} - \frac{f}{2}\right)\frac{\partial A(t_D)}{\partial t_D} = -k_{ro}^{max} J' f \varepsilon^{b+1-1/f} A(t_D)^{1/f} \tag{36}$$

Once again for Equation (36) to be finite in the limit $\varepsilon \to 0$, we require $f = 1/(1+b)$, Equation (24). Then:

$$\frac{1}{A(t_D)^{1+b}}\frac{\partial A(t_D)}{\partial t_D} = -\frac{2(2+b)}{b(3+2b)}k_{ro}^{max} J' \tag{37}$$

Integrating once:

$$A(t_D) = \frac{1}{(\eta t_D + C)^{1/b}} \tag{38}$$

where:

$$\eta = \frac{1+b}{3+2b}\beta = \frac{2(2+b)}{(3+2b)}k_{ro}^{max} J' \tag{39}$$

and C is a constant. We determine the constant by insisting that the average saturation is continuous at $t_D = t_{D1}$. From the early time solution, Equation (28) and the late time solution Equations (35) and (38) we find:

$$C = \left[\frac{b(3+2b)}{2(1+b)^2}\right]^b - \eta t_{D1} \tag{40}$$

Then we can write:

$$\bar{S}(t_D) = 1 - \frac{1+b}{(2+b)(1+\kappa(t_D - t_{D1}))^{1/b}} \tag{41}$$

where:

$$\kappa = \left[\frac{2(2+b)}{3+2b}\right]\left[\frac{2(1+b)^2}{b(3+2b)}\right]^b k_{ro}^{max} J' \tag{42}$$

As mentioned previously the solution is controlled by the boundary condition at the inlet ($x_D = 0$). The behavior is dominated by the insistence that the flow rates of oil and water are equal but in opposite directions as the oil mobility tends to zero. Note that unlike the empirical correlation, Equations (1)–(3), in our formulation the imbibition rate is independent of the water relative permeability and viscosity. However, the imbibition rate – given by $\kappa t_D / t$ – is proportional to the interfacial tension and inversely proportional to the length of the system squared, as demonstrated experimentally (Zhang et al., 1996; Morrow and Mason, 2001).

3. Comparison with Experiment

In this section we compare the early and late time predictions against experimental imbibition data on cores. The expressions for the average saturation use multiphase flow properties, namely the dimensionless capillary pressure gradient at $S = 1, J'$, the oil relative permeability exponent, b, and the maximum oil relative permeability, k_{ro}^{max}. In the experiments we study, the capillary pressure and relative permeability were not measured. In order to obtain genuine predictions obtained using independently measured experimental data, we will take the values of b and k_{ro}^{max} measured in steady state for waterflooding strongly water-wet Berea sandstone by Oak (1990) – in this case $k_{ro}^{max} = 1$ and the oil relative permeability varies approximately linearly with saturation, giving $b = 1$. While we do not have imbibition capillary pressure measurements from Oak, the relative permeability data is well predicted using pore-scale network modeling (Blunt et al., 2002, Valvatne and Blunt, 2004) where predictions of capillary pressure were also made. Using the network model the predictions of capillary pressure as function of water saturation are given for a strongly water-wet sandstone of porosity $\phi = 0.24$, permeability $K = 2554$ mD, and interfacial tension $\sigma = 30$ mN/m. We then use Equation (14) to find the dimensionless capillary pressure $J(S_w)$ and plot it against water saturation S_w in Figure 1. The gradient of the dimensionless capillary pressure at $S = 1, J'$, is then obtained by measuring the slope of the tangent at $S_w = 1 - S_{or}$, which is plotted in Figure 1. We then obtain $J' = 0.19$ (see also typical imbibition capillary pressures in Bear (1972) that give similar or larger values for J'

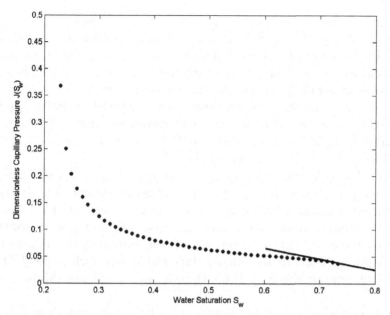

Figure 1. Dimensionless capillary pressure J versus water saturation S_w derived from a pore network model of Berea sandstone (Blunt et al., 2002; Valvatne and Blunt, 2004). The solid line is the tangent at $S_w = 1 - S_{or}$.

on sand and bead-packs). With these values we can write our predictions as follows. At early times $t_D < t_{D1}$, $\bar{S}(t_D)$ is given by Equation (28) with $\gamma = 0.063$, $t_{D1} = 1.75$. For late times $t_D > t_{D1}$, $\bar{S}(t_D)$ is given by Equation (41) with $\kappa = 0.36$.

To evaluate our theoretical solution, we use three different published experimental data on imbibition (Mattax and Kyte, 1962; Hamon and Vidal, 1986; Zhang *et al.*, 1996), in which the oil recovery was measured as a function of time. These data sets are for different porous media, core dimensions, boundary conditions, and oil and water viscosities. All of the systems were strongly water-wet. Mattax and Kyte (1962) performed experiments to test a scaling theory for imbibition oil recovery from fractured reservoirs. Two sets of experimental data were used: one set was for two cylindrical sandstone core samples of different sizes with all-faces-open (AFO) to imbibition. The other set was for four alundum cylindrical cores of different lengths with only one-end-open (OEO) to imbibition. The two sets of data had different oil/water viscosity ratios. The two sets, therefore, provided examples of boundary conditions, viscosity ratio, and porous media that were all different. Hamon and Vidal (1986) studied the effect of height and boundary conditions on recovery. The cores used were made of aluminum silicate with different lengths and boundary conditions of AFO, OEO and two-ends-open (TEO). The same water and oil phase with constant viscosity ratio were used in all the experiments. In order to test the applicability of Equation (2) an extensive experimental study was performed by Zhang et al. (1996). Imbibition measurements were presented for cylindrical Berea sandstone cores of different sizes, and different boundary conditions of AFO, TEO, OEO, and two-ends-closed (TEC). The viscosity of oil varied in these experiments. Ma et al. (1995, 1997) applied the modified scaling group to all the above-mentioned data on oil recovery (Mattax and Kyte, 1962; Hamon and Vidal, 1986; Zhang et al., 1996) and found a close correlation between them.

In Figure 2 we plot oil recovery as a function of the empirical dimensionless time t_{DE}, Equation (2), for the different experiments studied. Experiments with different boundary conditions, core sizes and fluid viscosities all plot on approximately the same universal curve that is reasonably well fitted by the empirical exponential function, Equation (1). The empirical fit is a good match to the Zhang data but is not such a good fit at early and late time to the data from Hamon and Vidal and Mattax and Kyte.

In this work for simplicity we considered a boundary condition of type OEO applied to a matrix block of length L. In order to extend the results to other boundary conditions of TEO, TEC and AFO, which were used in the experiments, we replace L by the modified characteristic length L_c,

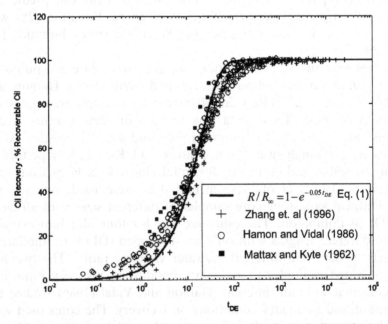

Figure 2. Experimental recovery data as a function of dimensionless time, t_{DE}, Equation (1) and an exponential empirical fit to the data, Equation (2).

Equation (3) in the equations for recovery and average saturation. This then enables us to compare our theory against all sets of experimental data (Mattax and Kyte, 1962; Hamon and Vidal, 1986; Zhang *et al.*, 1996) with different sample sizes and shapes and boundary conditions. Starting from the experimental data we obtain the results of oil recovery as a function of time, and plot them as a function of the dimensionless time, Equation (17). The results are shown in Figure 3, where for comparison we have also plotted the analytical oil recovery from Equations (28) and (41) with $t_{D1} = 1.75$, $\gamma = 0.063$ and $\kappa = 0.36$. The fraction of recoverable oil, plotted in Figure 3, is the same as the average normalized saturation, $\bar{S}$. The experimental data plot approximately onto one universal curve, although there is noticeably more scatter than in Figure 2, indicating that our dimensionless time t_D is not as good a scaling group as t_{DE} since we ignore any dependence on the water viscosity. However we obtain a good match between the theory and experiments for the majority of the data, especially those of Hamon and Vidal and Mattax and Kyte.

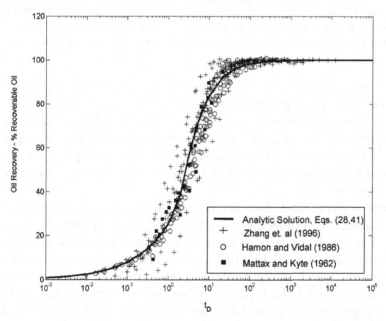

Figure 3. Experimental oil recovery data as a function of dimensionless time t_D, Equation (17) compared to analytical predictions from Equations (28) and (41). Notice that in comparison with Figure 2 there is slightly more scatter in the data, implying that the dimensionless time t_D, which ignores the water viscosity, is not such a good scaling group as t_{DE}. However, the match to the data by the analytical solution is superior to the empirical correlation, Equation (1), and involves no adjustable parameters.

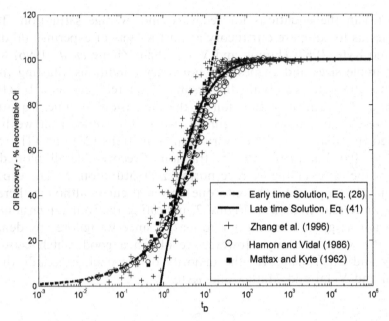

Figure 4. Experimental data compared with early, Equation (28), and late time, Equation (41), analytical recovery solutions versus dimensionless time t_D. The early-time solution over-predicts recovery at late times. The transition time t_{D1}, from early to late time solutions is where the dashed and solid lines first intersect.

Comparing Figures 2 and 3 it is evident that while Equation (2), that includes the water viscosity, provides a slightly better correlation than the dimensionless time we propose, Equation (17), our analytic predictions of recovery match the data significantly better than the empirical exponential correlation. Furthermore, our expressions are first-principles predictions based on independently measured data, while the empirical exponential form is a fit to the data.

The same experimental data are plotted in Figure 4, where early and late time analytical solutions are also plotted separately. Figure 4 shows the transition time t_{D1}, from early to late time solutions, where the dashed and solid lines first intersect. In Figure 4 we can also see that the square root of time behavior, which is represented by the early time solution, is not a good approximation for the recovery at late time in a finite-sized medium. Furthermore, extrapolating the late time solution to earlier times predicts negative recoveries for t_D less than around 1.

4. Conclusions

We have used the integral method to construct an approximate analytical expression for the water saturation during one-dimensional counter-current imbibition in a strongly water-wet medium. We assign a finite viscosity to the displaced phase (oil or NAPL) and show that the boundary condition at the inlet, where the displaced phase flow must be finite as its mobility tends to zero, controls the overall character of the solution.

The predictions of analytical expressions for early and late times are compared with several sets of experimental data, and with an empirical exponential form for the recovery proposed in the literature. Our expression fits the experimental data well and gives a better estimate of recovery than the exponential form without any adjustable parameters.

Acknowledgements

We would like to thank the sponsors of the ITF project on 'Improved Simulation of Flow in Fractured and Faulted Reservoirs' for supporting this research.

Appendix A. Solutions with an Infinite Capillary Pressure Gradient

Several empirical models – notably those proposed by van Genutchen (1980), Lehnard and Parker (1987a,b,c) and Parker et al. (1987) – assume that the capillary pressure gradient diverges at $S = 1$. We now consider this case. We write in analogy to Equation (15):

$$\left.\frac{\partial P_c}{\partial S_w}\right|_{S_w=1-S_{or}} = -\sigma\sqrt{\frac{\phi}{K}}J^*\varepsilon^{-c} = \frac{1}{(1-S_{wi}-S_{or})}\left.\frac{\partial P_c}{\partial S}\right|_{S=1} \tag{A.1}$$

where c is some exponent and J^* an appropriate pre-factor. Then Equation (18) becomes:

$$\frac{\partial \bar{S}}{\partial t_D} = k_{ro}^{max}J^*\varepsilon^{b-c}\left.\frac{\partial S}{\partial x_D}\right|_{x_D=0} \tag{A.2}$$

The analysis follows exactly as before with the same expressions for the saturation profile. The only difference is that to produce sensible solutions we require:

$$f = \frac{1}{1+b-c} \tag{A.3}$$

The expressions for average saturation etc are the same as before but with $b-c$ substituted for b and J^* for J'.

The van Genutchen (1980), Lehnard and Parker (1987a,b,c) and Parker et al. (1987) models use (in our notation):

$$P_c = nm^{1/n}\sigma J^*\sqrt{\frac{\phi}{K}}\left(1-S^{-1/m}\right)^{1/n} \tag{A.4}$$

where $m = 1 - 1/n$, and:

$$k_{ro} = (1 - S)^{1/2} \left(1 - S^{1/m}\right)^{2m} \tag{A.5}$$

Then taking $S = 1 - \varepsilon$ and taking the limit $\varepsilon \to 0$:

$$c = 1 - \frac{1}{n} \tag{A.6}$$

$$b = \frac{5}{2} - \frac{2}{n} \tag{A.7}$$

To be consistent with our analysis, we also need to use:

$$k_{ro}^{\max} = m^{-2m} \tag{A.8}$$

n, the exponent in the van Genutchen model (Dullien, 1992), has a value typically slightly greater than 1.

References

Aronofsky, J. S., Masse, L. and Natanson, S. G.: 1958, A model for the mechanism of oil recovery from the porous matrix due to water invasion in fractured reservoirs, *Trans. AIME* **213**, 17–14.

Barenblatt, G. I., Entov, V. M. and Ryzhik, V. M.: 1990, *Theory of Fluid Flows Through Natural Rocks*, Kluwer Academic Publishers, Dordrecht.

Barenblatt, G. I., Patzek, T. W. and Silin, D. I.: 2003, The Mathematical model of non-equilibrium effects in oil-water displacement, *SPE J.* **8**(4), 409–416.

Bear, J.: 1972, *Dynamics of Fluids in Porous Media*, Dover Publications Inc., New York.

Blunt, M. J., Jackson, M. D., Piri, M. and Valvatne, P. H.: 2002, Detailed physics, predictive capabilities and macroscopic consequences for pore-network models of multiphase flow, *Adv. in Water Resour.* **25**, 1069–1089.

Chen, J., Miller, M. A. and Sepehmoori, K.: 1995, Theoretical investigation of counter-current imbibition in fractured reservoir matrix blocks, in: *SPE 29141 Proceedings of the 13th SPE Symposium on Reservoir Simulation*, San Antonio, TX, USA, February 12–15, 1995.

Chen, Z. X., Bodvarson, G. S. and Witherspoon, P. A.: 1990, An Integral equation formulation for two-phase flow and other nonlinear flow problems through porous media, in: *SPE 20517 Proceedings of the SPE Annual Technical Conference and Exhibition*, New Orleans, September 23–26, 1990.

Dullien, F. A. L.: 1992, *Porous Media: Fluid Transport and Pore Structure*, 2nd edn, Academic Press Inc., San Diego.

Hamon, G. and Vidal, J. V.: 1986, Scaling up the capillary imbibition process from laboratory experiments on homogeneous heterogeneous samples, in: *SPE 15852, Proceedings of the Eur. Pet. Conf.*, London, October 20–22, 1986.

Kashchiev, D. and Firoozabadi, A.: 2003, Analytical Solutions for 1D Countercurrent Imbibition in Water-Wet Media, *SPE J.* **8**(4), 401–408.

Lenhard, R. J. and Parker, J. C.: 1987a, A model for hysteretic constitutive relations governing multiphase flow, 1. Saturation-pressure relations. *Water Resour. Res.* **23**(12), 2187–2196.

Lenhard, R. J. and Parker, J. C.: 1987b, A model for hysteretic constitutive relations governing multiphase flow, 2. Permeability-saturation relations. *Water Resour. Res.* **23**(12), 2197–2206.

Lenhard, R. J. and Parker, J. C.: 1987c, Measurement and prediction of saturation pressure relationships in three phase porous media systems, *J. Contaminant Hydrol.* **1**, 407–424.

Ma, S., Morrow, N. R. and Zhang, X.: 1997, Generalized scaling of spontaneous imbibition data for strongly water-wet systems, *J. Pet. Sci. Eng.* **18**, 165–178.

Ma, S., Zhang, X., and Morrow, N. R.: 1995, Influence of fluid viscosity on mass transfer between rock matrix and fractures, in: *Paper CIM 95–94 proceedings of the Pet. Soc. CIM 46th Annual Technical Meeting*, Banff, Alberta, May 14–17.

Mattax, C. C. and Kyte, J. R.: 1962, Imbibition oil recovery from fractured, water-drive reservoir, *SPE J.* 177–184.

McWhorter, D. B. and Sunada, D. K.: 1990, Exact integral solutions for two-phase flow, *Water Resour. Res.* **26**(3), 399–413.

Morrow, M. R. and Mason, G.: 2001, Recovery of oil by spontaneous imbibition, *Curr. Opin. Colloid Interface Sci.* **6**, 321–337.

Oak, M. J.: 1990, Three-phase relative permeability of water-wet Berea, in: *SPE 20183, Proceedings of the SPE/DOE Seventh Symposium on Enhanced Oil Recovery*, Tulsa.

Parker, J. C., Lenhard, R. J. and Kuppusamy, T.: 1987, A parametric model for constitutive properties governing multiphase flow in porous media. *Water Resour. Res.* **23**(4), 618–624.

Pooladi-Darvish, M. and Firoozabadi, A.: 2000, Cocurrent and countercurrent imbibition in a water-wet matrix block, *SPE J.* **5**(1), 3–11.

Rangel-German E. R. and Kovscek A. R.: 2002, Experimental and analytical study of multidimensional imbibition in fractured porous media, *J. Pet. Sci. Eng.* **36**, 45–60.

Reis, J. C. and Cil, M.: 1993, A model for oil expulsion by counter-current water imbibition in rocks: one-dimensional geometry, *J. Pet. Sci. Eng.* **10**, 97–107.

Valvatne, P. H. and Blunt, M. J.: 2004, Predictive pore-scale modeling of two-phase flow in mixed wet media, *Water Resour. Res.* **40**, W07406, doi:10.1029/2003WR002627.

van Genuchten, M. T.: 1980, A closed form equation for predicting the hydraulic conductivity of unsaturated soils, *Soil Sci. Soc. Am. J.* **44**, 892–898.

Zhang, X., Morrow, N. R. and Ma, S.: 1996, Experimental verification of a modified scaling group for spontaneous imbibition, *SPE Reserv. Eng.* **11**, 280–285.

Zhou D., Jia, L., Kamath, J. and Kovscek, A. R.: 2002, Scaling of counter-current imbibition processes in low-permeability porous media, *J. Petrol. Sci. Eng.* **33**, 61–74.

Zimmerman, R. W. and Bodvarsson, G. S.: 1991, A simple approximate solution for horizontal infiltration in a Brooks-Corey Medium, *Trans. Porous Media* **6**, 195–205.

Zimmerman, R. W. and Bodvarsson, G. S.: 1989, An approximate solution for one-dimensional absorption in unsaturated porous media, *Water Resour. Res* **25**, 1422–1428.

Zimmerman, R. W., Bodvarson, G. S. and Kwicklis, E. M.: 1990, Absorption of water into porous blocks of various shapes and sizes, *Water Resour. Res.* **26**(11), 2797–2806.

Transp Porous Med (2005) 58:191–216
DOI 10.1007/s11242-004-5501-5

Multi-Stage Upscaling: Selection of Suitable Methods

G. E. PICKUP[1,*], K. D. STEPHEN[1], J. MA[1], P. ZHANG[1] and
J. D. CLARK[1,2]
[1]*Institute of Petroleum Engineering, Heriot-Watt University, Riccarton, Edinburgh, EH14
4AS, Scotland, UK*
[2]*Now with ChevronTexaco, 6001 Bollinger Canyon Road, Building D-1180, San Ramon,
CA 94583, USA*

(Received: 14 August 2003; in final form: 24 March 2004)

Abstract. Reservoirs are often composed of an assortment of rock types giving rise to permeability heterogeneities at a variety of length-scales. To predict fluid flow at the full-field scale, it is necessary to be aware of these different types of heterogeneity, to recognise which are likely to have important effects on fluid flow, and to capture them by upscaling. In fact, we may require a series of stages of upscaling to go from small-scales (mm or cm) to a full-field model. When there are two (or more) phases present, we also need to know how these heterogeneities interact with fluid forces (capillary, viscous and gravity). We discuss how these effects may be taken into account by upscaling. This study focusses on the effects of steady-state upscaling for viscous-dominated floods and tests carried out on a range of 2D models are described. Upscaling errors are shown to be reduced slightly by the increase in numerical dispersion at the coarse scale. We select a combination of three different upscaling methods, and apply this approach to a model of a North Sea oil reservoir in a deep marine environment. Six different genetic units (rock types) were identified, including channel sandstone and inter-bedded sandstone and mudstone. These units were modelled using different approaches, depending on the nature of the heterogeneities. Our results show that the importance of small-scale heterogeneity depends on the large-scale distribution of the rock types. Upscaling may not be worthwhile in sparsely distributed genetic units. However, it is important in the dominant rock type, especially if there is good connectivity through the unit between the injector wells (or aquifer) and the producer wells.

Key words: permeability heterogeneity, permeability upscaling, balance of forces

1. Introduction

Heterogeneities in permeable rocks exist over a wide range of length-scales, from mm upwards. Often these heterogeneities arise in a hierarchical fashion: for example, laminae occur at the mm–cm scale, and sets of laminae combine to form beds at dm–m scales (e.g. Weber, 1986; Jones *et al.*, 1993;

---

*Author for correspondence: e-mail: gillian.pickup@pet.hw.ac.uk

Kjonsvik *et al.*, 1994; Pickup and Hern, 2002). For multi-phase flow, in addition to the effects of permeability structure, we need to consider the interaction of small-scale heterogeneity with capillary pressure. Laboratory experiments (Huang *et al.*, 1995), and numerical simulation (Ringrose *et al.*, 1993; Van Lingen *et al.*, 1996, 1997) of water floods in oil reservoirs have shown that capillary trapping may occur between laminae, leading to reduced oil recovery.

A multi-stage upscaling approach, called the Geopseudo Method, was developed to take account of small-scale structures in reservoir-scale models (Corbett *et al.*, 1992; Pickup *et al.*, 2000). In this method, the upscaling stages are defined by the predominant length-scales in the reservoir rocks (e.g. the typical bed size). However, upscaling from small-scale models increases the time taken for reservoir simulation, and we do not wish to use it unless we know that the small-scale structures are important. Ringrose *et al.* (1999) developed a set of guidelines based on parameters such as permeability contrast and layer thickness, to estimate when small-scale structures could be important. Jones *et al.* (1993) and Kjonsvik *et al.* (1994) investigated the effects of heterogeneity in hierarchical models. They found that large-scale structures had a dominant effect, but small-scale structures were also important.

In this paper, we investigate the applicability of various methods for upscaling two-phase flow, and demonstrate how multi-stage upscaling may be carried out in a practical way in a real reservoir model, and we provide further insights into the importance of upscaling small structures. Brief reviews of the effects of different types of heterogeneities and of different upscaling methods are presented in Section 2, followed by a discussion on the choice of upscaling method (Section 3). In Section 4, we describe some numerical tests which were performed to investigate errors in steady-state upscaling. Then we introduce the field study, and describe the multi-stage upscaling (Sections 5 and 6) procedure. The upscaling results are presented in Sections 7 and 8, followed by a summary and conclusions in Section 9.

2. Description of Upscaling Methods

A number of people have written reviews on upscaling methods, for example: Renard and Marsily (1997), Barker and Thibeau (1997) and Christie (2001). In this paper, we do not wish to present a review of upscaling methods. We focus on oil-water systems, and we describe the two-phase approaches which are used here.

2.1. STEADY-STATE METHODS

When two phases are present (e.g. oil and water), we need to upscale relative permeability in addition to absolute permeability. The easiest approach to two-phase upscaling is to assume that the fluids are in steady-state, i.e. $\partial S_w / \partial t = 0$, where S_w is water saturation. This converts a two-phase problem into two single-phase problems (ignoring gravity):

$$\nabla \cdot (k_{ro} \, k_{abs} \, \nabla P_o) = 0 \qquad \nabla \cdot (k_{rw} \, k_{abs} \, \nabla P_w) = 0, \tag{1}$$

where k_{abs} is the absolute permeability tensor, k_r is the relative permeability, P is pressure and the subscripts o and w refer to oil and water, respectively. Steady-state methods are fast, and also have the advantage that they produce "well-behaved" effective relative permeability functions (Ekrann and Aasen, 2000).

Steady-state upscaling may be carried out easily when a flood is in capillary equilibrium (negligible viscous forces, i.e. negligible injection rate), or when it is in viscous-dominated steady-state (negligible capillary forces). (Note, also that when the flood is gravity-dominated, the fluids may reach vertical equilibrium, which is also a steady-state.) Steady-state upscaling methods have been discussed by a number of authors, e.g. Smith, 1991; Kumar and Jerrauld, 1996; Pickup and Stephen, 2000.

When a viscous-dominated flood is in steady-state

$$u_t \cdot \nabla f_w = 0, \tag{2}$$

where u_t is the total flow (oil and water), and f_w is the fractional flow of water. This is equivalent to the fractional flow being constant along streamlines (Ekrann and Aasen, 2000). In this study, we assume that the fractional flow is constant on the inlet face of a model. For a system with isotropic relative permeabilities, this means that the fractional flow is constant throughout the model.

2.2. TWO-PHASE DYNAMIC

This is the most difficult type of upscaling. It is time-consuming, because a two-phase flow simulation is required (albeit on a local fine grid). Also, the results may not be reliable (Barker and Thibeau, 1997). There are several advantages in two-phase dynamic upscaling:

1. A fine-scale two-phase flow simulation is used to reproduce the water saturation distribution for the correct balance of forces, provided appropriate boundary conditions are used.
2. The dispersion of the front due to permeability heterogeneities, or capillary or gravitational forces is taken into account.
3. The upscaled relative permeabilities (pseudos) can compensate for the increase in numerical dispersion as the grid is coarsened.

We used the Kyte and Berry (1975) method in this study. In this method, the total flow of each fluid is calculated at the outlet face of a coarse block, by summing the fine-scale flows. Then the average pressure of each phase is computed from the values in the central column (plane in 3D) of the coarse block. Kyte and Berry (1975) used the product of the fine cell height, the permeability and the relative permeability as a weighting factor. The relative permeabilities are then computed using Darcy's law.

3. Choice of Upscaling Method

When carrying out multi-stage upscaling, we need to choose a method or methods which will provide reasonable results, but also we need to ensure that the methods are feasible. For example, we are not likely to choose to carry out two-phase dynamic upscaling at each level, because it would be too time consuming. There are several factors to be considered, and these are discussed below.

3.1. PERMEABILITY STRUCTURE AND CONTRAST

For single-phase flow, it has been shown that upscaling is robust when we have a large separation of length-scales (e.g. Whitaker, 1969). Often this condition does not hold, leading to inaccuracies in upscaling. (Errors may be reduced to a certain extent, however, using non-uniform upscaling e.g. Durlofsky *et al.*, 1997). Permeability averaging may be used for upscaling models with simple structures, such as continuous parallel layers, or correlated random fields (assuming the correlation length is much smaller than the size of a coarse grid cell). Often averaging, or a combination of averages (arithmetic–harmonic, or harmonic–arithmetic) is used for speed in more complex models (e.g. Li *et al.*, 2001). When the permeability variation is small (permeability contrast 5:1 or less), the errors will not be severe (Pickup and Hern, 2002). However, large permeability contrasts are frequently encountered in oil reservoirs, and complex structures such as cross-bedded rocks, in which case the pressure solution method may produce more accurate results.

3.2. BALANCE OF FORCES

We are concerned with two-phase flow in this study (oil and water), and we need to take account of the balance of viscous, capillary and gravity forces, which determine the paths of fluids through a reservoir. We assume here that oil (the non-wetting phase) is being displaced by water (the wetting phase). Upscaling is easiest when one force dominates, and the system reaches a steady-state, and we consider the effect of each force in turn.

3.2.1. *Viscous-Dominated Flow*

In a viscous-dominated flood, the injected fluid will tend to travel faster in the high permeability regions, leading to a dispersion of the flood front (e.g. Zhang and Tchelepi, 1999; Neuweiler *et al.*, 2003). See Figure 1. Artus *et al.* (2004) and Noetinger *et al.* (2004) take account of viscous cross-flow between high and low permeability regions in layered and isotropic models, respectively. They show that the amount of dispersion depends on the mobility ratio at the flood front, defined as:

$$M_f = \frac{\lambda(S_f)}{\lambda(S_{wc})}. \tag{3}$$

λ is the total mobility, defined as

$$\lambda = \lambda_o + \lambda_w = k_{ro}/\mu_o + k_{rw}/\mu_w, \tag{4}$$

where μ is viscosity. S_f is the saturation at the Buckley–Leverett shock front (Buckley and Leverett, 1942) and S_{wc} is the connate (minimum) water saturation. Artus *et al.* (2004) show that, for $M_f < 1$ (stable flood), fluctuations in the front due to heterogeneities become stationary, whereas if $M_f > 1$ (unstable flood), perturbations in the front grow. Noetinger *et al.* (2004) show that this is also true for isotropic stochastic media. Zhang and Tchelepi (1999) show that the amount of dispersion in a correlated random model increased with the standard deviation and correlation length of the permeability field.

Many engineers only perform single-phase upscaling in two-phase systems (e.g. Christie and Blunt, 2001), in which case the dispersion in the front due to heterogeneities will not be taken into account. On the other hand, in a finite-difference flow simulation, the flood front is spread out by the finite size of the grid cells, giving rise to numerical dispersion (Lantz, 1971). In the case of immiscible flow, this is not a true dispersion, since the width of the front does not grow with $\sqrt{t}$. The front is self-sharpening, and tends to a constant width (Hewett and Behrens, 1991). Therefore, if only single-phase upscaling is performed, numerical dispersion will tend

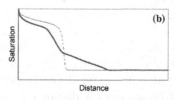

Figure 1. (a) Example of a saturation distribution in a heterogeneous medium water (dark colour) is being injected into the left side of the model. (b) The average saturation profile (solid line), compared with the shock front from a homogeneous medium.

to compensate for loss of physical dispersion in a stable flood. (This effect has been reported by Haajizadeh et al. (2000) for miscible floods.)

Ekrann and Aasen (2000) show that the viscous-dominated steady-state method is most likely to apply for stable displacements and small-scale heterogeneities (i.e. a correlated random permeability distribution with a correlation length much smaller than the coarse grid cell). Obviously the viscous-dominated steady-state method will suffer from the same inaccuracies as single-phase upscaling, in that it will not take account of the dispersion of the front due to heterogeneities. We describe additional tests carried out on steady-state upscaling in Section 4.

3.2.2. Capillary-Dominated Flow

In this discussion, we assume that the rock is water-wet or mixed-wet, and that the capillary pressure functions are similar to those used in the field study (Section 5, Figure 11). Capillary pressure is high at low water saturations, so water is imbibed into the oil-saturated zones, leading to a dispersion of the flood front in a homogeneous medium. In a heterogeneous medium, since capillary pressure varies as $\sqrt{\phi/k}$, water is imbibed from high permeability regions into low permeability ones, and so capillary pressure can help to stabilise the flood front. When a flood is capillary-dominated, the imbibition is so strong that water travels preferentially through the low permeability regions, leaving oil trapped in the high permeability zones (e.g. Ringrose et al., 1993; Huang et al., 1995).

At small-scales, the capillary-equilibrium upscaling method is often applied. For a homogeneous model, this method will be reasonably accurate if capillary pressure spreads out the front so that the water saturation varies negligibly over the length a coarse grid cell. A number of people have assessed the effect of capillary pressure using capillary numbers which describe the relative strength of the capillary pressure gradient compared with the viscous pressure gradient. Although the balance of forces varies throughout a flood (Stephen et al., 2001), capillary numbers are useful for approximate calculations. Yortsos and Fokas (1983) define a capillary number as:

$$B = \frac{k(dP_c/dS)_{char}}{vL\mu_o},$$
(5)

where v is flow rate, $(dP_c/dS)_{char}$ is a characteristic pressure gradient, and L is the length of the system. Ekrann et al. (1996) used this capillary number to estimate that, for a typical North Sea sandstone ($k = 1$ D), fluids are in capillary equilibrium over a distance of about 20 cm.

Tests of the capillary-equilibrium method, on a range of heterogeneous models, were described in Pickup and Stephen (2000). The method was not as accurate as dynamic two-phase upscaling, but performed well. The accuracy depends on the type of permeability structure. In models where high permeability regions are enclosed by low permeability regions, so that oil trapping can occur, the residual oil saturation is very sensitive to the capillary number. Using the capillary equilibrium method where it does not apply, will over-estimate the amount of oil trapping.

Often the capillary equilibrium method is applied over regions which are a few metres rather than a few centimetres across. However, this will not have much impact on the results of a full reservoir simulation, where the grid cells are commonly 100 m long.

3.2.3. Gravity-Dominated

In a gravity-dominated flood, the fluids segregate rapidly, with the low density fluid flowing above the high density one. In this case, vertical equilibrium may be assumed, and a 3D simulation may be reduced to 2D. Since the effect of gravity depends on the density difference between the fluids, gas floods are more likely to be gravity-dominated than water floods. We do not describe this method here, since we are focussing on oil–water systems. However, water slumping is frequently found in simulations of waterfloods. The amount of slumping depends on the vertical permeability distribution: if permeability increases downwards, the effect of slumping will be greater than if it increases upwards.

3.2.4. Intermediate Balance of Forces

Often a flood will not be dominated by a single force, so a combination of forces will have to be taken into account. Steady-state methods may still be used, but are more difficult to apply. For example Dale et al. (1997) consider steady-state upscaling for 1D heterogeneous media, with a range of viscous and capillary forces. Also, an intermediate viscous-capillary steady-state method was developed by Stephen and Pickup (2000), which can capture capillary trapping more accurately in a flood with an intermediate force balance. In general, though, when there is a mixture of forces, two-phase dynamic upscaling is required to reproduce the flood at the coarse-scale.

3.2.5. Balance of Forces Summary

In summary, permeability heterogeneities have different effects on a flood, depending on the balance of forces. Focussing on viscous and capillary forces: the flood front will move faster in the high permeability regions when the flood is viscous-dominated, but will advance faster in the low permeability regions when the flood is capillary-dominated. At an intermediate viscous-capillary ratio, the front will advance at the same rate in high and low permeability regions. The precise nature of the fluid distribution will depend on the permeability structure and contrast, and the balance of viscous:capillary forces. At small scales, water floods tend to be capillary-dominated and the capillary equilibrium method is useful for the first level of a multi-stage upscaling process. At larger scales, we need to consider the results of numerical tests before deciding which methods to use.

4. Numerical Tests

For full-field simulation, grid cells of around 100 m (in the horizontal) are often used, and the effects of the grid size on the simulation results must be considered. In this section we present the results of tests carried out to assess the effect of various errors, including numerical dispersion, on scale-up results.

The analytical approaches for estimating the effect of heterogeneities on two-phase flow which were discussed above (e.g. Zhang and Techelepi, 1999), assume that the permeability variation is in the form of small fluctuations about a mean value. In nature, however, permeability can vary by several orders of magnitude over a small distance. We have conducted a number of numerical experiments in order to examine upscaling in a range of 2D models, including highly heterogeneous models. The object of these tests was to examine the accuracy of performing single-phase only, or steady-state upscaling in a two-phase system (water flooding).

Examples of these models are shown in Figure 2. We assume that the permeability is a diagonal, isotropic tensor. Also, we assume that capillary forces have been taken into account at a smaller scale, and we focus here on viscous-dominated floods.

4.1. CORRELATED RANDOM MODELS

The first set of tests was performed on horizontal, isotropically correlated random models (Figure 2(a)). The permeability distribution was log-nor-

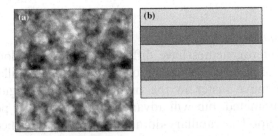

Figure 2. Examples of models used for testing upscaling: (a) correlated random model (horizontal plane and (b) layered model (vertical plane). In both cases darker shades represent lower permeability.

mal, and the relative permeabilities were given by:

$$k_{rw} = S_n^2 \qquad k_{ro} = (1 - S_n)^2, \tag{6}$$

where S_n is saturation normalised to go from 0 to 1. Table 1 lists the details of the models and the values of parameters used. All combinations of these parameters were tested. In the first set of simulations, a horizontal injector was placed on the left side of the model and a horizontal producer on the right side. In the second set, a quarter 5-spot pattern was modelled, with the injector and producer at opposite corners. We refer to these cases as the linear and diagonal models, respectively. Note that the linear flow case could represent flow confined to a high permeability channel, or alternatively could represent vertical flow of water from an aquifer at the bottom of the model to a horizontal well across the top. A number of scale-up factors was tested, as indicated in Table 1. In each case, the error in the coarse-scale simulation compared to the fine-scale simulation was calculated follows:

$$\text{RMS} = \sqrt{\frac{\sum\limits_{i=1}^{n}(R_{ci} - R_{fi})^2}{n}} \div \max(R_{fi}) \times 100\% \tag{7}$$

where R is the cumulative recovery. The subscripts c and f refer to "coarse" and "fine", and $i = 1, 2, \ldots, n$ refers to time steps.

The most obvious trend in the errors was that they increased with the scale-up factor (Figure 3). This is partly due to the error in single-phase upscaling increasing with the scale-up factor, but also the numerical dispersion at the coarse-scale model increases with scale-up factor. Apart from this the results of the two sets of simulations (linear and diagonal) were different.

Table I. Values of parameters used in the correlated random models

Parameter	Values
Grid size (no. of cells)	64 × 64
Cell dimensions (m)	3.05 × 3.05 × 3.05
Mean of ln (k)	5.3 (corresponds to 200 mD)
Std. deviation in ln (k)	0.1, 0.5, 1.0, 1.5, 2.0, 2.5, 3.0, 3.5
Correlation length (in terms of total model size)	0.0, 0.1, 0.2, 0.3
No. of realisations	8
Mobility ratio at flood front, M_f	0.6, 1.4
Scale-up ratio	2 × 2, 4 × 4, 8 × 8, 16 × 16

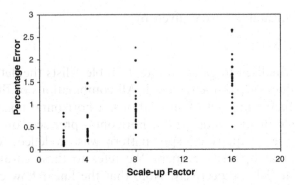

Figure 3. Percentage error in cumulative oil recovery, for the diagonal flow model, stable flood. The points represent different values of the standard deviation and correlation length.

4.1.1. *Linear Flow*

In the linear flow case, the upscaling errors tended to decrease and then increase as the standard deviation and correlation length were increased, as shown in Figure 4(a) and (b). This is because the errors in the coarse-scale simulation are due partly to the loss of fine-scale physical dispersion, and partly due to numerical dispersion in the coarse-scale simulation. At intermediate levels of heterogeneity (intermediate values of the standard deviation and correlation length), the numerical dispersion partly compensates for the loss of physical dispersion, and this reduces the errors. The errors were larger for the unstable flood, but the trend was similar.

4.1.2. *Diagonal Flow*

In the case of diagonal flow, breakthrough at the production well is less sharp, and the effect of numerical dispersion in the coarse-grid simulations is less noticeable. This means that as the amount of small-scale physical dispersion increases (larger standard deviation and correlation length, unstable flood), the errors in the coarse-scale simulation tend to increase. Certainly the errors were much larger for unstable floods when the standard deviation was large ($\sigma > 1.5$). Contour plots of the percentage error are shown in Figure 4(c) and (d) for the case where the scale-up factor was 4×4. The errors tend to be larger for an intermediate value of correlation length ($L = 0.1$). This is because the coarse-scale blocks are similar in size to the correlation length. (As mentioned previously upscaling is only accurate when there is a wide separation of length scales, so the coarse grid block should be much larger or much smaller than the correlation length.)

4.2. HORIZONTAL LAYERS

The layered models consisted of five layers of alternating high and low permeability (high, low, high, low high). See Figure 2(b). Three models were

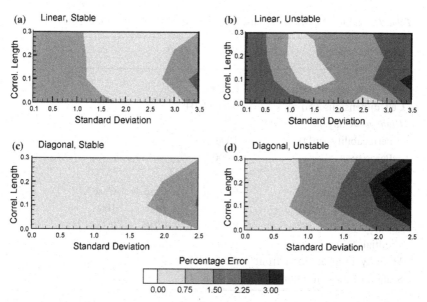

Figure 4. Percentage error in cumulative oil recovery in the correlated random models, for a scale-up factor of 4×4. (The maximum standard deviation plotted for the diagonal case is 2.5, because there was a large variation in the errors for cases with larger standard deviations.)

created with different properties in the low permeability layer, referred to below as models a, b, and c. The details of the models are described in Table II. The values of the porosity and initial water saturation were chosen to be similar to the field study described in Section 5. (The initial water saturation increased with decreasing permeability). Water was injected at the left and oil produced at the right. The densities of oil and water were set similar to minimise gravity effects. In each case, the models were upscaled to a 1D model (using the arithmetic average). Different sizes for the coarse-scale cells were used, as shown in Table II. The errors in the coarse-scale simulations were again calculated using Equation 7. Figure 5 shows the errors for the stable case. The main effect is that the errors increase as the permeability contrast between the layers decreases. This is because porosity varies with permeability, and the initial water saturation increases with decreasing permeability. Therefore, there is very little oil in model a, which almost behaves as if it were homogeneous. As the permeability in the low permeability layers increases, the effects of dispersion are more noticeable, and the upscaling errors increase. However, as the scale-up factor increases, the upscaling errors in cases b and c decrease slightly. This is because numerical dispersion is compensating for physical dispersion. The simulations with an unstable mobility ratio gave similar results, but with slightly larger errors.

Table II. Value of parameters used in the horizontal layer models

Parameter	Values		
Grid size (no. of cells)	64×64		
Cell dimensions (m)	0.04×0.04		
High perm. layers			
Permeability (mD)	1000		
Porosity	0.300		
Initial S_w	0.10		
Low perm. layers	Case a	Case b	Case c
Permeability (mD)	1	10	100
Porosity	0.130	0.185	0.255
Initial S_w	0.70	0.60	0.25
Mobility ratio at flood front	0.53, 1.17		
Scale-up ratios in x-direction	2, 4, 8, 16		

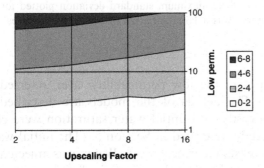

Figure 5. Percentage errors in cumulative oil recovery in the horizontal layer models.

The results from these two types of model indicate the nature of the errors likely to arise when upscaling a reservoir model. They suggest that the errors in cumulative recovery, when upscaling by a small factor, will not be severe, and that numerical dispersion will partly compensate for loss of physical dispersion. (Note, however, that the average water saturation profile for a simulation with fine-scale heterogeneity causing dispersion will not, in general, be the same as the profile for a coarse-scale simulation with numerical dispersion). The error from using only single-phase upscaling will depend on factors discussed above, namely: permeability contrast (standard deviation), correlation length, boundary conditions, and mobility ratio. In a 3D model, however, the percentage errors may be larger, because there is likely to be more physical dispersion (the fluid will have access to more high permeability pathways). Also, if the relative permeability curves and their end points vary, this may cause larger errors.

In the tests described above, the relative permeability curves had the same shape in every rock type. This means that the streamlines for two-phase flow are the same as those for single-phase flow (Ekrann and Aasen, 2000), and so single-phase upscaling produces the same results as the viscous-dominated steady-state upscaling. Therefore the results of these tests can be used directly to assess the accuracy of the viscous-dominated steady-state method.

4.3. 1D UPSCALING TESTS

Ideally, we would like to test the accuracy of a multi-stage upscaling procedure by comparing results at the finest and coarsest scales. However, it is impossible to carry out simulations on a fine enough grid, unless we use 1D models. We performed a series of upscaling tests in a 1D homogeneous model to assess which combination of methods steady-state and dynamic methods is both feasible and reasonably accurate. The model was horizontal, because there is usually a larger scale-up factor in the horizontal than in the vertical. The size of cells used in the test was influenced by the length scales in the field study, which is described below. The fine-scale model consisted of cells which were 5 cm long (about the size of a core plug). There were 10,000 of these cells, giving a total length of 500 m for the model. (The length of the reservoir studied was about 5 km.) The coarse-scale cells were 1 m for the first stage of scale-up, 25 m for the second stage, and 100 m for the final stage. The permeability of the model was set to 1200 mD and the porosity to 0.3. The injection rate was set to give an interstitial velocity of 30 cm per day.

We applied the upscaling methods in every combination at each scale. The water-cut and normalised oil recovery curves were examined for each coarse-scale model, and compared with the fine-scale result. As expected, some of the upscaling cases under-compensated for numerical dispersion, and some over-compensated (Figure 6).

One of the best cases used the following combination of methods:

- Stage 1 – steady-state (capillary equilibrium)
- Stage 2 – Kyte and Berry (1975)
- Stage 3 – steady-state (viscous-dominated)

We chose this approach in our upscaling study.

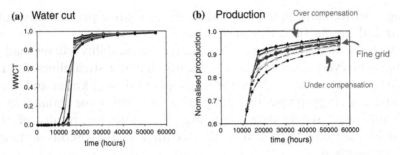

Figure 6. Comparison of water cut and production in the upscaling tests.

5. The Field Study

We have applied multi-level upscaling to a field in the UK central North Sea, which is about 5 km long by 2 km wide. The sediments were deposited in a channelised turbidite system, and comprise stacked sand-prone channels, over-bank inter-channel facies, and mud-rich debris flows. A detailed analysis of core data was carried out, and 6 genetic units (GU) were identified:

1. Channel sandstone
2. Argillaceous sandstone (containing laminae and mud clasts)
3. Interbedded sandstone and mudstone
4. Mudstone with injected sandstone
5. Slumps/debris flows
6. Mudstone

Figure 7 shows the relative proportions of the genetic units, and Table 3 lists the average porosities and permeabilities.

These units are typical of the types of structures which are found in turbidite environments, which act as oil reservoirs in the North Sea and in many other parts of the world (e.g. Gulf of Mexico, off-shore West Africa). Although all reservoirs are different, similar rock types exist in other turbidite reservoirs, so our results are relevant to a large number of oil reservoirs.

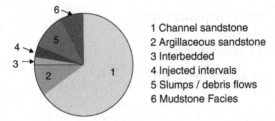

Figure 7. Relative proportions of the genetic units in the model.

Table III. Distribution of porosities and permeabilities

Genetic Unit	Porosity	Permeability		
		Average of ln(k)	Equivalent perm (mD)	Std. Dev. of ln(k)
1	0.288	7.0	1060	3.5
2	0.255	5.1	168	1.5
3–6	0.184	3.7	42	2.3

A "fine-scale" stochastic geological model was generated using the Roxar RMS package, and for convenience, we refer to this model as the "RMS model". A sector of this model, which was used to test upscaling is shown in Figure 8. The cells in the RMS model were $25\,m \times 25\,m \times 1.5\,m$. The horizontal size was taken from the resolution of the seismic data, and the vertical size was chosen from an analysis of the thicknesses of the facies. The stochastic modelling was carried out in two stages. First genetic units 2, 3, 4 and 6 were modelled as flat ellipsoidal bodies to form the background facies. The thicknesses and widths were modelled stochastically using uniform or triangular distributions. The thicknesses ranged from 1.5 to 5 m, and widths from 50 to 1000 m. The lengths were 1.5 times the widths. (The distributions were slightly different for each unit.) The dimensions of the bodies were chosen from outcrop studies, and the relative proportions were derived from cores (Figure 7). Genetic units 1 and 5 were then modelled as channels, the thickness of which varied from about 1.5 to 4 m, and the widths were approximately 100 times the thickness. The orientations of all the bodies were distributed uniformly between east

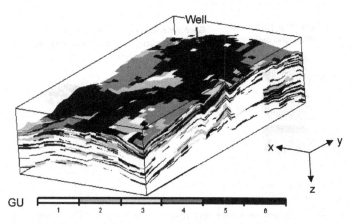

Figure 8. Sector of the RMS model. North is along the *y*-axis.

and south. The model consisted of approximately 4 million active cells. The petrophysical properties of the model were assigned stochastically, and were constrained to well log data. The porosities were derived from well logs, and the permeabilities for the base-case model were estimated from core-plug k–ϕ relationships. The permeability distribution in the model therefore consisted of several log-normal distributions, as shown in Table III. A fuller account of the geological interpretation and modelling is presented in Stephen *et al.*, 2002.

As can be seen from Figure 7 and Table III, 75% of the model consisted of high and intermediate permeability channel sandstone (units 1 and 2). The remainder of the model was made up of ellipsoidal bodies a few metres thick and a few hundreds of metres across. Thus, although this study is specific to one reservoir, the results can be applied to reservoirs with similar structure.

The cells of the RMS model were relatively coarse compared with the size of the fine-scale structures within each genetic unit. Small-scale models were therefore created as follows, according to the geological modelling hierarchy shown in Figure 9. The number of cells in the geological model was too large for two-phase flow simulation, so the RMS model had to be coarsened for full-field simulation.

GU1 consisted mainly of high permeability sand, but there was also some inter-bedded sandstone and mudstone. As a base case, we assumed that GU1 contained only homogeneous, high permeability sandstone. However, using the results of outcrop studies, we also included the inter-bedded facies (GU3) in this unit as part of a sensitivity study (Section 7).

GU2 contained poorer quality sandstone compared to GU1, with mud clasts which were modelled stochastically as shown in Figure 10(a).

GU3 consisted of laterally continuous layers of sandstone and mudstone, and was modelled stochastically, as shown in Figure 10(b). These

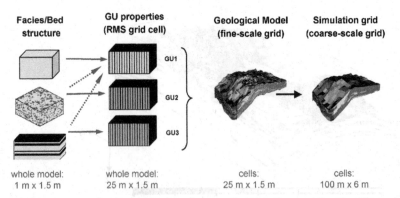

Figure 9. Schematic diagram of scales required for the models. The dimensions indicate sizes in the horizontal and vertical directions.

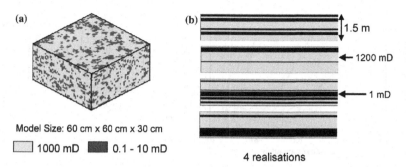

Figure 10. (a) Model used for GU2 (one realisation). (b) Model used for GU4 (4 realisations).

models are similar to the horizontal layer models, case a, described in Section 4.

GU4 contained mainly mudstone, but with sandstone dykes and sills. Sensitivity studies described in Stephen *et al.* (2002) showed that the injected sandstone structures had a negligible effect, so in this paper, the fine-scale structure in GU4 has been ignored.

GU5 and GU6 were low permeability, non-reservoir facies, and were not modelled at the fine scale.

The field was produced by 4 horizontal wells, and pressure support was provided by an aquifer. The densities of the fluids were $0.862\,g/cm^3$ for oil and $1.02\,g/cm^3$ for water (atmospheric conditions), and the viscosities were $0.81\,cP$ for oil and $0.44\,cP$ for water. (These values varied with pressure).

The SCAL data for this field was unreliable, so we generated Corey-type synthetic relative permeability curves with an exponent of 2 (Brooks and Corey, 1964):

$$k_{ro} = (1 - S_n)^2 \qquad k_{rw} = 0.35 S_n^2. \tag{8}$$

The flood was stable, with a mobility ratio at the front of $M_f = 0.5$. Note that in a real field study, there is not much point in spending a lot of time with two-phase upscaling if the input data are unreliable. However, in this work, we were more interested in studying the sensitivities, rather than in producing an accurate prediction of recovery. The effect of using alterative relative permeability curves in this field was investigated by Di Pierro *et al.*, 2003.

The average connate water saturation for each genetic unit, and for pure mudstone was derived from the well logs. The residual oil was assumed to be constant at 0.25. An oil–water capillary pressure curve for sandstone was derived from air–brine drainage curve (by scaling). This curve was converted into an imbibition curve for a mixed-wet rock by adding a negative "tail". The capillary pressure curves for the other genetic units were formed

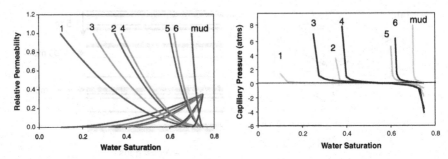

Figure 11. Input relative permeability and Pc curves. The numbers refer to the genetic units, and "mud" refers to the curves used for pure mudstone in the small-scale models.

by scaling the sandstone curve by $\sqrt{\phi/k}$, using the average porosity (ϕ) and average permeability (k) for each unit. Figure 11 shows the input curves used for each genetic unit.

6. Application of Upscaling Methods

6.1. STAGE 1 – ACCOUNTING FOR THE SMALL-SCALE STRUCTURE

We performed upscaling on the models of small-scale structures within GU2 and GU3. First single-phase upscaling was performed. The pressure solution method, with no-flow boundary conditions, was used for GU2. Since the models for GU3 were essentially 1D, they were upscaled using the arithmetic average for along-layer flow and the harmonic average for across-layer. The results of the single-phase upscaling were used to estimate the kv/kh ratio (ratio of vertical to horizontal permeability) for those units in the cells of the RMS model.

Next, two-phase upscaling was carried out on the GU2 and GU3 models. The models were upscaled using the capillary-equilibrium method to obtain effective relative permeabilities. In the case of GU2, the up-scaled curves were similar to the average of the input curves, because the fluid flowed mainly through the sandstone, avoiding the mud clasts. However, for GU3, the resulting effective permeabilities were highly anisotropic (Figure 12(a)). This was because, for vertical flow, oil became trapped in the high permeability sandstone layers upstream of the low permeability mudstone. (See, for example, Huang *et al.*, 1995). The effective connate water saturation was different in each realisation of the model (due to the different amounts of sandstone and mudstone). However, when the water saturations were normalised, the curves were all similar (Figure 12(a)). When applying these relative permeabilities, we averaged the normalised curves, so that only a single curve was required for all GU3, and we used the average S_{wc} from well logs.

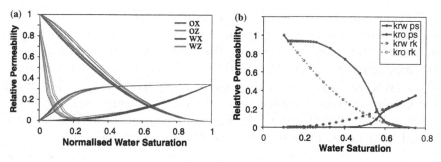

Figure 12. (a) Normalised effective relative permeability curves for GU3. (b) Pseudo relative permeabilitie for GU1 (solid lines). The input curves are shown using dashed lines.

The relative permeabilities for the other units were assigned an exponent of 2 (as the input curves), with the connate water saturation set to the average from the well logs, for that unit. The residual oil was again set to 0.25.

6.2. STAGE 2 – UPSCALING TO THE RMS MODEL CELL

Since the genetic units were laterally extensive, we assumed that each cell of the RMS model contained only one type of genetic unit, so that each $25\,m \times 25\,m \times 1.5\,m$ cell, consisted of 25×25 identical sub-units. Scale-up in this case was only applied in the horizontal direction, and was only used to reduce the numerical dispersion in the final simulation. The Kyte and Berry (1975) method was used to calculate the pseudo relative permeabilities for each genetic unit. An example of the results are shown in Figure 12(b), for GU1.

6.3. STAGE 3 – UPSCALING THE RMS MODEL

The full-field RMS model contained over 4 million cells, and we reduced these by a factor of 64 (4 in each direction), using single-phase upscaling (pressure solution method in RMS). We also tested Stage 3 upscaling, using the viscous-dominated steady-state method, on a sector model which was small enough for us to perform a simulation on the RMS-scale grid. The sector model (Figure 8) consisted of $24 \times 44 \times 64$ cells, which were upscaled to $6 \times 11 \times 16\,(= 1056)$ cells. A regular cartesian grid was used, because the upscaling software used this type of grid.

7. Comparison of Results

Simulations for sensitivity studies were carried out on the sector model (Figure 8). This model contained one horizontal well and pressure support was provided by an aquifer, which was modelled using the Fetkovitch

(1971) method. Since the well was controlled by oil rate (as observed in the field), the models were compared by plotting the water cut (Figure 13) and the bottom hole pressure as a function of time (not shown). Note that the production schedule used in the simulation was the same as that used in the real field, and includes periods when the well was shut in, giving the graphs an irregular appearance. In the following cases, the RMS grid (25 m × 25 m × 1.5 m) was used, except in the fourth model which was up-scaled by a factor of 4 in each direction (as described above):

(1) Case 1 – the base case model – ignores fine-scale structure. Also ignores capillary pressure (set to zero), and has $S_{wc} = 0.1$ for all genetic units.
(2) Case 1a – the base case permeabilities, but with a different S_{wc} for each unit, and capillary pressure.
(3) Case 2 – Case 1a upscaled to account for structure in GU2 and GU3 (Stage 1).
(4) Case 3 – includes Kyte and Berry upscaling for Stage 2.
(5) Case 4 – accounts for small-scale structure (Case 2) and includes Stage 3 upscaling.
(6) Case 5 – similar to Case 2, but also accounting for heterogeneities in GU1.

The results shown in Figure 13 indicate that modifying the base case (Case 1) to take account of capillary pressure (Case 1a) has a significant effect on the water cut. This is because capillary pressure produces a transition zone between the water and the oil, allowing earlier breakthrough of water. Upscaling to account for fine-scale structure in GU2 and GU3 had little effect (Case 2). This is not surprising because GU2 and GU3 make up a small fraction of the volume.

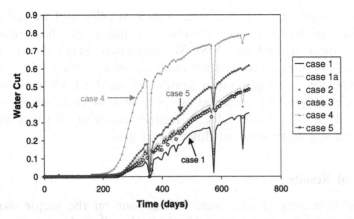

Figure 13. Comparison of water cuts in the sector model. Note that the curves for Cases 1a, 2 and 3 are almost superimposed.

The Kyte and Berry upscaling stage had a negligible effect here: the curves for Cases 2 and 3 in Figure 13 are almost super-imposed. This is because the flow was mainly in the vertical direction from the aquifer, whereas the Kyte and Berry upscaling affected flow in the horizontal direction.

There is a significant difference between Case 4, which included the final upscaling, and Case 2, which used the fine-scale RMS grid. Some of this discrepancy could have been caused by the re-sampling of the RMS grid to form a regular Cartesian grid for the upscaling package. However, the main problem is that the single-phase upscaling at this stage is inaccurate, due to the high permeability contrasts and the size of the heterogeneities. Another problem is that the top of the aquifer is only 28.65 m (19 RMS layers) below the top of the model. Therefore there are only a few coarse grid layers between the aquifer and the horizontal well, so the upscaling gives a very early breakthrough. The errors in the third stage of upscaling were reduced when we considered a version of the model with vertical wells (Section 8).

The reservoir model was dominated by GU1 (sandstone channels), which was modelled initially as homogeneous, high permeability material, and so most of the fluid therefore flowed through GU1. However, GU1 did contain some heterogeneities in the form mudstone layers (as in GU3). Therefore, for Case 5, we constructed a simple model consisting of sandstone (GU1) with a layer of inter-bedded material (GU3), and upscaled this using the capillary equilibrium method. The resulting effective relative permeabilities (which were similar to those in Figure 12(a)) were used in the sector model, in place of the GU1 relative permeabilities. Figure 13 shows that the heterogeneities in GU1 have a much larger effect than those in GU2 and GU3, because GU1 was much more prevalent than GUs 2 and 3 (Figure 7).

8. Vertical Well Model

The reservoir studied here had horizontal wells and pressure support was provided by an aquifer. This meant that the flow was mainly vertical. In this section, we investigate the effect of vertical wells (horizontal flow). We changed the sector model so that an injection well was located at coordinates (7,7), and a production well at (18,38). Both wells were completed in all the layers. The injector well was controlled by rate (which was constant), and the producer was controlled by bottom-hole pressure. Figure 14 (a) shows a comparison of the water cut for several models with vertical wells. As above, Case 1 is the base case, Case 2 contains small-scale upscaling, and Case 3 also has Kyte and Berry upscaling. There is a significant difference between the Case 1 and Case 2 but, as explained in Section 7, this is manly due to the addition of capillary pressure. This time, there is a slight difference between Cases 2 and 3, due to the effect of compensation for numerical dispersion by the Kyte and Berry method.

We also compared Cases 2 and 4 using vertical wells, to investigate the accuracy of the stage 3 upscaling (viscous-dominated-steady-state). (Once again this level of upscaling was carried out using the results from stage 1, omitting the Kyte and Berry intermediate stage, which had little effect.) Figure 14(b) shows that in the vertical well model, this stage of upscaling is more accurate than in the original model, although there is still a significant error. The error in cumulative oil recovery due to the third level of upscaling (assuming the Case 2 results are "correct") is 3.5%.

The error in the third level of upscaling (vertical well case) can be compared with the results for upscaling stochastic models, described in Section 4. For the sector model, the over-all standard deviation in $\ln(k)$ was 3.7, and the flood was stable. The correlation lengths of the permeability distribution were approximately $Lx = Ly = 0.5$, and $Lz = 0.16$ (in terms of the model length). The correlation length is therefore larger than in the tests, where the maximum correlation length was $Lx = 0.3$. The error in recovery for the test models, with $\sigma = 3.5$ and $Lx = 0.3$, was about 1%, which is smaller than the error in the field model. However, the field model is a 3D model, rather than a 2D one, and we would expect small-scale dispersion to be greater in 3D, so errors in upscaling would be larger.

9. Summary and Conclusions

In this study, we have investigated the feasibility of multi-stage upscaling. The applicability of a range of different methods was discussed. In particular we considered steady-state upscaling, because it is quicker, and is therefore feasible. We presented the results of tests of steady-state upscaling in viscous-dominated systems. The errors caused by upscaling are due to the loss of fine-scale dispersion, which depends on the model parameters (standard deviation and correlation length), and also on factors such as the

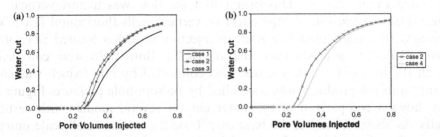

Figure 14. Comparison of water cuts for the simulations with vertical wells. (a) Cases 1, 2 and 3. (b) Cases 2 and 4.

mobility ratio and the flow regime (linear or diagonal). However, numerical dispersion can partly compensate for the loss of physical dispersion. This reduces the errors, particularly where there is linear flow.

We then applied multi-stage upscaling to a real field model. The three stages of upscaling were:

1. The capillary equilibrium method
2. The Kyte and Berry (1975) method
3. The viscous-dominated steady-state method

The results of sensitivity studies showed that the effect of small-scale structure was not really significant in the case of GU2 and GU3, because these units are not very prevalent, and occur as small, disconnected bodies. Most of the difference between the base case and Case 2 is caused by assigning a different S_{wc} for each genetic unit and applying capillary pressure (Case 1a). However, small-scale heterogeneity in GU1 is important, because it makes up about 2/3 of the model. The intermediate level Kyte and Berry upscaling stage did not make any difference in the field case, where the flow was mainly vertical. When vertical wells were used for injection and production, so that the flow was mainly horizontal, the Kyte and Berry upscaling had a slight effect. However, it was small compared to the errors in the third level of upscaling. (Compare Figures 14 (a) and (b)). Therefore in this study, the intermediate stage of upscaling was unnecessary.

The error in the third stage of upscaling (which was the only level which could be checked) was larger than predicted by the tests in Section 4. In the original model (aquifer and horizontal producer), the error was very large. This was due to the proximity of the aquifer to the well. (To improve the results, we should perhaps have coarsened the grid more in the aquifer, and retained the fine grid at the top of the reservoir). In the vertical well case, the results were more satisfactory, although the error was still larger than expected. However, the tests were conducted on 2D models whereas the field study used a 3D model.

In summary, the most important level of upscaling in this study was stage 3. Stage 2 was unnecessary. Stage 1 was also unnecessary in the base case, but was important when heterogeneities were included in GU1, the most prevalent rock type. This upscaling approach will also be viable in other reservoirs containing a similar distribution of length scales (not necessarily just turbidite reservoirs). In other words, it is applicable to reservoirs where there are two separate levels of heterogeneity: a small scale (cm–dm) and a much larger scale (100s m horizontally and a few metres vertically).

This study highlights the fact that the importance of small-scale structure cannot be judged by examining only small-scale models. Previous studies of the effect of heterogeneity have tended to assume that the same type of heterogeneity extends across a reservoir (e.g. Van Lingen *et al.*, 1997; Pickup and Hern, 2002), in which case, small-scale structure will have a significant effect. However, in reservoirs where only some of the units contain small-scale heterogeneity, the large-scale connectivity of the high permeability units is obviously important. If there are high permeability pathways between injector wells and producer wells (or between an aquifer and producing wells), most of the fluid will follow these pathways rather than flowing through lower permeability, heterogeneous units. This suggests that upscaling studies should begin with an examination of the large-scale structure and then focus on small-scale structure if it is likely to have a significant effect.

Acknowledgements

This research was carried out as parts of the Genetic Units, Heterogeneity and Upscaling Projects at Heriot-Watt University. We should like to thank the following companies for their sponsorship: Anadarko, BG, BP, Chevron, Conoco, the UK DTI, Enterprise, ExxonMobil, JNOC, Petrobras, Shell, Statoil, Unocal and Veba. We thank Roxar for the use of the RMS modelling package and Schlumberger for the use of the Eclipse flow simulation package. We should also like to thank Mauricio Silva for his help in analysing the data and generating the RMS models.

References

Artus, V., Noetinger, B. and Ricard, L.: 2004, Dynamics of the water-oil front for two-phase, immiscible flows in heterogeneous porous media. 1 – stratified media, *Transport Porous Media* **56**, 283–303.

Barker, J. W. and Thibeau, S.: 1997, A critical review of the use of pseudorelative permeabilities for upscaling, *SPE Reservoir Eng.* **12**, 138–143.

Brooks, R. H. and Corey, A. T.: 1964, Hydraulic properties of porous media, Hydrology Paper 3, Civil Engineering Department, Colorado State University, Fort Collins.

Buckley, S. E. and Leverett, M. C.: 1942, Mechanism of fluid displacement in sands, *Trans. AIME* **146**, 107.

Christie, M. A.: 2001, Flow in porous media – scale up of multiphase flow, *Curr. Opin. Colloid Interface Sci.* **6**, 236–241.

Christie, M. A. and Blunt, M. J.: 2001, Tenth SPE comparative solution project: a comparison of upscaling techniques, *SPE Reservoir Eval. Eng.* **4**(4), 308–317.

Corbett, P. W. M., Ringrose, P. S., Jensen, J. L. and Sorbie, K. S.: 1992, laminated clastic reservoirs: the interplay of capillary pressure and sedimentary architecture, SPE 24699, presented at the 67th SPE Annual Technical Conference, Washington, 4–7 October 1992.

Dale, M., Ekrann, S., Mykkeltveit, J. and Virnovsky, G.: 1997, Effective relative permeabilities and capillary pressure for one-dimensional heterogeneous porous media, *Transport Porous Media* **26**, 229–260.

Di Pierro, E., Stephen, K. D., McDougall, S. R. and Pickup, G. E.: 2003, Sensitivity analysis on the two-phase flow properties in a turbidite reservoir, SPE 81020, presented at the

SPE Latin American and Caribbean Petroleum Engineering Conference, Port-of-Spain, Trinidad, West Indies, 27–30 April 2003.

Durlofsky, L. J., Jones, R. C. and Milliken, W. J.: 1997, A non-uniform coarsening approach for the scale up of displacement processes in heterogeneous porous media, *Adv. Water Resour.* **20**, 335–347.

Ekrann, S. and Aasen, J. O.: 2000, Steady-state upscaling, *Transport Porous Media* **41**(3), 245–262.

Ekrann, S., Dale, M., Langaas, K. and Mykkeltveit, J.: 1996, Capillary limit effective two-phase properties for 3D media, SPE 35491, proceedings of the European 3-D Reservoir Modelling Conference, Stavanger, Norway, 16–17 April, 1996.

Fetkovitch. M. J.: 1971, A Simplified approach to water influx calculations – finite aquifer systems, *J. Petroleum Technol.* 814–828.

Haajizadeh, M., Fayers, F. J. and Cockin, A. P.: 2000, Effects of phase behaviour, dispersion and gridding on sweep patterns for nearly miscible gas displacement, SPE 62995, presented at the SPE Annual Technical Conference, Dallas, Texas, 1–4 October, 2000.

Hewett, T. A. and Behrens, R. A.: 1991, Scaling laws in reservoir simulation and their use in a hybrid finite difference/streamtube approach to simulating the effects of permeability heterogeneity, in: L. W. Lake, H. B. Carroll and T. C. Wesson (eds), *Reservoir Characterization II*, Academic Press, San Diego, California, pp. 402–441.

Huang, Y., Ringrose, P. S. and Sorbie, K. S.: 1995, Capillary trapping mechanisms in water-wet laminated rocks, *SPE Reservoir Eng.* **10**(4), 287–292.

Jones, A., Doyle, J., Jacobsen, T. and Kjonsvik, D.: 1993, Which sub-seismic heterogeneities influence waterflood performance? A case study of a low net-to-gross fluvial reservoir, presented at the 7th European IOR Symposium, Moscow, Russia, October 27–29, 1993.

Kjonsvik, D., Doyle, J., Jacobsen, T. and Jones, A.: 1994, The effects of sedimentary heterogeneities on production from a shallow marine reservoir – what really matters?, SPE 28445, presented at the European Petroleum Conference, London, UK, 25–27 October, 1994.

Kumar, A. T. and Jerauld, G. R.: 1996, Impact of scale-up on fluid flow, from plug to grid-block scale in reservoir rock, SPE 35452, presented at the SPE/DOE Tenth Symposium on Improved Oil Recovery, Tulsa, 21–24 April 1996.

Kyte, J. R. and Berry, D. W.: 1975, New pseudo functions to control numerical dispersion, *SPE J.* 269–275.

Lantz, R. B.: 1971, Quantitative evaluation of numerical diffusion (truncation error), *SPE J.* 315–320.

Li, D., Becker, B. and Kumar, A.: 2001, A new efficient averaging technique for scaleup in multimillion-cell geologic models, *SPE Reservoir Eval. Eng.* **4**(4), 297–307.

Neuweiler, I., Attinger, S., Kinzelbach, W. and King, P.: 2003, Large scale mixing for immiscible displacement in heterogeneous porous media, *Transport Porous Media* **51**, 287–314.

Noetinger, B., Artus, V. and Ricard, L.: 2004, Dynamics of the water-oil front for two-phase, immiscible flows in heterogeneous porous media. 2 – isotropic media, *Transport Porous Media* **56**, 305–328.

Pickup, G. E. and Stephen, K. D.: 2000, As assessment of steady-state scale-up for small-scale geological models, *Petroleum Geosci.* **6**, 203–210.

Pickup, G. E., Ringrose, P. S. and Sharif, A.: 2000, Steady-state upscaling: from lamina-scale to full-field model, *SPE J.* **5**(2), 208–217.

Pickup, G. E. and Hern, C. Y.: 2002, The development of appropriate upscaling procedures, *Transport Porous Media* **46**, 119–136.

Renard, P. and Marsily, G. de: 1997, Calculating equivalent permeability: a review, *Adv. Water Resour.* **20**(5–6), 253–278.

Ringrose, P. S., Sorbie, K. S., Corbett, P. W. M. and Jensen, J. L.: 1993, Immiscible flow behaviour in laminated and cross-bedded sandstones, *J. Pet. Sci. Eng.* **9**(2), 103–124.

Ringrose, P. S., Pickup, G. E., Jensen, J. L. and Forrester, M. M.: 1999, The ardross reservoir gridblock analog: sedimentology, statistical representivity, and flow upscaling, in: R. Schatzinger and J. Jordan, (eds), *Reservoir Characterization – Recent Advances*, AAPG Memoir **71**, 265–276.

Smith, E. H.: 1991, The influence of small-scale heterogeneity on average Relatirve permeability, in: L. W. Lake, H. B. Carroll and T. C. Watson (eds), *Reservoir Characterization II*, Academic Press, Inc.

Stephen, K. D. and Pickup, G. E.: 2000, A fully-implicit upscaling method for accurate representation of the balance of viscous and capillary forces, presented at the 7th European Conference on the Mathematics of Oil Recovery, Baveno, Italy, 5–8 September 2000.

Stephen, K. D., Pickup, G. E. and Sorbie, K. S.: 2001, The local analysis of changing force balances in immiscible incompressible two-phase flow, *Transport Porous Media* **45**, 63–68.

Stephen, K. D., Clark, J. D. and Pickup, G. E.: 2002, Modelling and flow simulations of a north sea turbidite reservoir: sensitivities and upscaling, SPE 78292, presented at the SPE 13th European Petroleum Conference, Aberdeen, Scotland, UK, 29–31 October, 2002.

Van Lingen, P. P, Bruining, J. and van Kruijsdijk, C. P. J. W.: 1996, Capillary entrapment caused by small scale wettability heterogeneities, *SPE Reservoir Eng.* 93–99.

Van Lingen, P. P. and Knight, S.: 1997, Evaluating capillary trapping within reservoir flow units, SPE38934, Proceedings of the SPE Annual Technical Conference, San Antonio, Texas, 5–8 October 1997, 941–956.

Weber, K. J.: 1986, How Heterogeneity Affects Oil Recovery, in L. W. Lake, and H. B. Carroll, (eds), *Reservoir Characterization*, Academic Press, Orlando, Florida.

Whitaker, S.: 1969, Advances in the theory of fluid motion in porous media, *Ind. Chem.* **12**, 14–28.

Yortsos, Y. C. and Fokas, A. S.: 1983, An analytical solution for linear waterflood including the effects of capillary pressure, *SPE J.* **23**(1), 115–124.

Zhang, D. and Tchelepi, H.: 1999, Stochastic analysis of immiscible two-phase flow in heterogeneous media, *SPE J.* **4**, 380–388.

DYNAMIC EFFECTS IN MULTIPHASE FLOW: A PORE-SCALE NETWORK APPROACH

T. GIELEN[1,*], S.M. HASSANIZADEH[2], A. LEIJNSE[3]
and H.F. NORDHAUG[4]
[1]*Faculty of Civil Engineering and Geosciences, Delft University of Technology, P.O. Box 5048, 2600 GA Delft, The Netherlands*
[2]*Environmental Hydrogeology Section, Utrecht University, P.O. Box 80021, 3508 TA Utrecht, The Netherlands*
[3]*Netherlands Institute for Applied Geosciences (NITG-TNO), P.O. Box 80015, 3508 TA Utrecht, The Netherlands*
[4]*Department of Mathematics, University of Bergen, Johannes Brunsgate 12, 5008 Bergen, Norway*

Abstract. Current theories of multiphase flow rely on capillary pressure and saturation relationships that are commonly measured under static conditions. To incorporate transient behaviour, new multiphase flow theories have been proposed. These include an extended capillary pressure-saturation relationship that is valid under dynamic conditions. In this relationship, the difference between the two fluid pressures is called dynamic capillary pressure, and is assumed to be a function of saturation and its time rate of change. The dependency is through a so-called damping coefficient. In this work, this proportionality between capillary pressure and saturation rate of change is investigated using a pore-scale network model. It consist of a three-dimensional network of tubes (pore throats) connected to each other by pore bodies. The pore bodies are spheres and pore throats are cylinders. Numerical experiments are performed wherein typical experimental procedures for both static and dynamic measurements of capillary pressure-saturation curves are simulated. The value of the damping coefficient is determined for one realisation of our network model. Then, the effect of different averaging domains on capillary pressure-saturation curves is investigated.

1. Introduction

Traditionally, multiphase flow in porous media is studied using pressure cells, in which quasi-static displacement experiments are performed. In these experiments, also referred to as multistep outflow experiments, fluid flow is controlled by capillary forces. The difference between the pressures of the fluids across the inflow and outflow boundaries is increased in small steps each time equilibrium between fluid phases has been achieved. These pressure steps are shown as the dashed-dotted line in figure 1. The equilibrium points are shown as the grey symbols. From these equilibrium points, the capillary pressure-saturation curve is then constructed (the dotted line). However, following this procedure, the dynamics of the flow inbetween the equilibria are not captured. The capillary pressure-saturation

---

* e-mail: t.w.j.gielen@citg.tudelft.nl

D.B. Das and S.M. Hassanizadeh (eds.) Upscaling Multiphase Flow in Porous Media, 217–236
© 2005 Springer. Printed in the Netherlands

relationship one would expect to find when including the flow dynamics is shown as the solid line in figure 1. This is illustrated in the paper by Dahle et al. (this book).

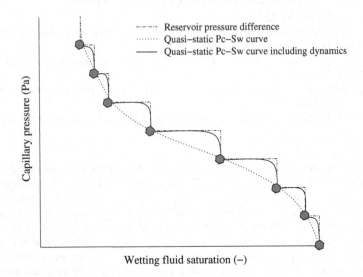

Figure 1. Determination of a capillary pressure-saturation curve. The dashed-dotted line shows the boundary pressure differences. The grey symbols denote the equilibrium points, from which the quasi-static $P^c - S^w$ curve is determined (the dotted line). The solid line shows the $P^c - S^w$ curve one would expect to find when flow dynamics are not discarded.

As early as 1921, one has realised that the behaviour of capillary flow under (quasi-)static and transient conditions is different [13]. In the 1960's and 1970's, several researchers found different capillary pressure-saturation relationships for static, steady-state and transient experiments [12, 10, 11]. An overview of the experiments is given by Hassanizadeh et al. [8].

To account for the observed dynamic effects in multiphase flow, an extended definition was proposed by Hassanizadeh & Gray, where the traditional (quasi-) static relationship is replaced by [6, 7]:

$$P^n - P^w = P^c_{stat} - \tau \frac{dS^w}{dt} = P^c_{dyn} \qquad (1)$$

where P^c_{stat} is traditional capillary pressure, P^c_{dyn} is dynamic capillary pressure, S^w is wetting fluid saturation, P^n and P^w are nonwetting and wetting fluid pressures, respectively, and τ is a damping coefficient that can be a function of saturation.

From experimental data published in literature, some average values for τ have been calculated by Hassanizadeh et al. [8]. However, a detailed study of equation (1), the proportionality between $P^n - P^w$ and $\frac{\partial S_w}{\partial t}$, and the dependence of τ on saturation has not been performed yet. Experimental studies to cover these issues will be very expensive and time consuming. Therefore, for a preliminary study of equation (1), we develop and employ a pore-scale network model.

Network models consist of an assembly of tubes as pore throats and pore bodies connecting them. These models can be used as a numerical surrogate for laboratory experiments, where every step and every parameter can be controlled and/or determined.

Numerous pore-scale network models have been developed in recent years. They are mostly static pore-scale network models, which are used to study quasi-static displacements. The dynamics of flow are not computed. Based on the boundary conditions and the curvature of the fluid-fluid interface, equilibrium positions are found. These are then used to update saturations and capillary pressure. Time is of no importance in these models. Boundary conditions similar to those of traditional laboratory experiments are imposed. For an overview, see Dullien [5] or Celia et al. [3].

Another type of pore-scale network models is the so-called dynamic network models. In these models, fluid-fluid interfaces are tracked in time and the flow of the fluids is simulated explicitly. Clearly, the interfaces do not have to be at equilibrium positions. Instead of increasing inflow boundary fluid pressure in small steps to a maximum value, the simulations are carried out with the boundary pressure set equal to a large constant value from the start, see, e.g. Mogensen & Stenby [9], Dahle & Celia [4] and Celia et al. [2]. Alternatively, one can impose a constant nonwetting flux at the boundary (e.g. Blunt & King [1]). In contrast to quasi-static pore-scale network models, dynamic models are generally very CPU-time consuming.

In this paper, first the dynamic pore-scale network model is described, stating the used assumptions and equations. Next, the numerical experiments to perform are described. The value of damping coefficient τ is computed for one realisation of our pore-scale network. Then, the effect of averaging domain size on capillary pressure-saturation curves is investigated. The results are discussed, and finally, conclusions are given.

2. The Dynamic Network Model

This section describes our dynamic pore-scale network model. This model is basically similar to the model of Blunt & King [1]. The main difference is that Blunt & King used Neumann conditions at the nonwetting phase inflow boundary. In our model, we have employed Dirichlet conditions, which enables us to simulate quasi-static drainage experiments as well. For this purpose, algorithms have been included for determining equilibrium between fluid phases after incremental increase of boundary conditions. Also, while we focus on dynamic effects in capillary pressure-saturation curvers, Blunt & King mainly focus on relative permeabilities.

2.1. ASSUMPTIONS

The pore network is a regular lattice of pore bodies and pore throats. A pore body is spherical in shape, a pore throat is a cylinder (see figure 2). Pore body and throat sizes are assumed to follow a truncated log-normal distribution. The following simplifying assumptions are made in this model:

- Both fluids are considered to be incompressible; the solid matrix is rigid.

- The pore throats have negligible volume, but they offer hydraulic resistance to flow.

- The passage of a fluid-fluid interface through a pore throat occurs instantaneously (i.e. the residence time is negligible compared to the characteristic flow time).

- Only one fluid resides in a pore throat at a time.

- Flow in pore throats is laminar and its rate is given by Poiseuille's formula.

- Local capillary pressure is zero for all pore bodies. Thus, there is only one pressure at a pore body.

- Gravity is neglected, hence fluid flow is driven by pressure gradients only.

From these assumptions, it follows that only one fluid pressure field for the whole network has to be solved at any given time. Nevertheless, there is a local-scale capillary pressure originating from pore throats entry pressure. It is equal to the difference in pressures of nonwetting and wetting phases occupying pore bodies adjacent to a given pore throat.

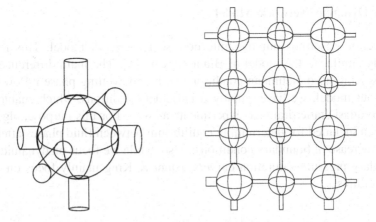

Figure 2. Schematic representation of a pore body and adjacent pore throats in a 3D network with coordination number 6 (left) and a 2D network with coordination number 4 (right).

2.2. GOVERNING PORE-SCALE EQUATIONS

In this section, pore-scale equations that are used in our network model under the assumptions mentioned in section 2.1 are described. Subscript i denotes pore body i, j denotes its neighbouring pore bodies, and ij the connecting pore throats. Superscript w and n denote wetting and nonwetting fluid phases, respectively.

Continuity equations
In the absence of external sources and sinks, the principle of volume conservation for pore body i connected to neighbouring pore bodies j ($j \in N_i$) can be written as:

$$V_i \frac{\Delta S_i^\alpha}{\Delta t} + \sum_{j \in N_i} Q_{ij}^\alpha = 0, \qquad (\alpha = w, n) \tag{2}$$

where V_i is volume, t is time, S_i^α is local α-phase saturation (fraction of volume V_i filled by fluid α), Q_{ij}^α is α-phase flux from pore body i to j, N_i is number of pore bodies connected to i.

Summation of equation (2) for both phases results in the following system of equations:

$$\sum_{j \in N_i} Q_{ij}^w + Q_{ij}^n = 0 \tag{3}$$

This follows from restriction:

$$S_i^w + S_i^n \equiv 1 \tag{4}$$

Flux equations
The flow of each phase within pore throat ij is assumed to be given by Poiseuille's formula. Wetting and nonwetting phase fluxes are expressed in terms of wetting and nonwetting phase pressures (neglecting gravitational terms):

$$Q_{ij}^\alpha = K_{ij}^\alpha \left(P_j^\alpha - P_i^\alpha \right), \qquad (\alpha = w \text{ or } n) \tag{5}$$

where K_{ij}^α is α-phase conductivity, P_i^α and P_j^α are α-phase pressures in pore body i and j, respectively. K_{ij}^α depends on the fluid occupancy of pore throat ij:

$$K_{ij}^\alpha = \frac{\pi R_{ij}^4}{8 \mu^\alpha L_{ij}} \tag{6}$$

where R_{ij} is radius, L_{ij} is length, and μ^α is α-phase dynamic viscosity. For a given pore throat, equation (5) will be written either for $\alpha = w$ or $\alpha = n$ depending on the phase occupancy of that throat at that instance.

Final system of linear equations
Equations (3) and (5) can be combined to obtain a set of linear equations in terms of P. For a given pore body i, the typical equation reads:

$$\sum_{j\in N_i}\left[\left(K_{ij}^w + K_{ij}^n\right)\left(P_i - P_j\right)\right] = 0 \qquad (7)$$

Only one pressure field, rather than both fluid phase pressure fields has to be solved. The summation of conductivities is allowed since only one fluid can occupy a pore throat at a time. When wetting phase occupies the pore throat, non-wetting phase conductivity K_{ij}^n equals zero, and vice versa.

2.3. NUMERICAL SOLUTION PROCEDURE

The governing equations presented in the previous section are solved for fluid pressure P and wetting fluid saturation S^w using an IMPES (IMplicit Pressure - Explicit Saturation) algorithm. Equation (7) is solved implicitly for P, assuming that S^w and throat conductances K^α are known from the previous time step. Then, equation (5) is used to calculate fluid fluxes. The fluid fluxes are substituted in equations (2) and local phase saturations are calculated. The procedure is then repeated. The flow chart describing the steps taken in the pore-scale network model is given in figure 3. To solve the set of equations (7), a Preconditioned Conjugate Gradient Method (PCGM) algorithm is used.

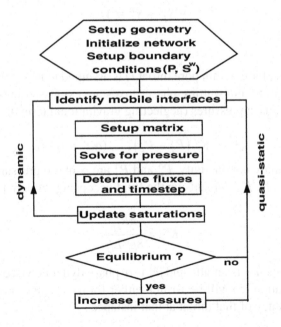

Figure 3. Flow chart of the pore-scale network model. Simulations stop when maximum time or maximum number of time steps have been exceeded (dynamic), or when final equilibrium has been achieved (quasi-static).

2.4. AVERAGING PROCEDURE

Pore-scale variables are averaged to arrive at the macro scale, where the results can be interpreted. The macro-scale pressures are defined as:

$$\langle P^\alpha \rangle = \frac{1}{\delta V^\alpha} \int_{\delta V^\alpha} P^\alpha \, dV = \frac{\sum_i^{N_n} P_i^\alpha S_i^\alpha V_i}{\sum_i^{N_n} S_i^\alpha V_i} \tag{8a}$$

$$\langle P^c \rangle = \langle P^n \rangle - \langle P^w \rangle \tag{8b}$$

where N_n is number of pore bodies, P_i^α is α-phase fluid pressure, S_i^α is α-phase saturation and V_i is volume. Average wetting and nonwetting fluid saturations are defined as:

$$\langle S^w \rangle = \frac{V^w}{V^n + V^w} = \frac{\sum_i^{N_n} S_i^w V_i}{\sum_i^{N_n} V_i} \tag{9a}$$

$$\langle S^n \rangle = 1 - \langle S^w \rangle \tag{9b}$$

For enhanced readability, the angular brackets are discarded in the remaining text. From this point onward, average wetting fluid saturation is simply referred to as saturation.

Volume averaging is performed over different domains in the network. The averaging domains are always in the center of the network. Their size is determined by the number of layers from the top and bottom of the network that will be discarded in the averaging procedure. Let the network size be nx·ny·nz, then the number of pore bodies N_n in the averaging domain is nx·ny·(nz-2·nskip). This is sketched in figure 4.

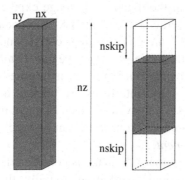

Figure 4. Illustration of different averaging domains in the network model, controlled by the amount of layers to be skipped in the averaging procedure (nskip).

3. Numerical Experiments for Determination of Dynamic Coefficient

3.1. PORE-NETWORK SIMULATION OF TYPICAL DRAINAGE EXPERIMENTS

The pore-scale network model is used to simulate typical laboratory experiments for the measurement of capillary pressure-saturation relationships and calculation of damping coefficient τ. As such, the pore-scale network is a surrogate for a sample of a porous medium (see figure 5). The nonwetting fluid reservoir is at the top of the network, the wetting fluid reservoir at the bottom.

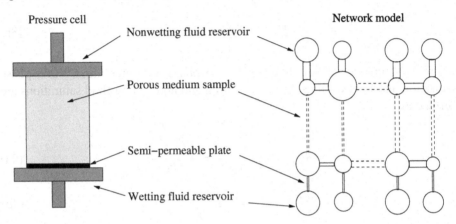

Figure 5. Schematic laboratory pressure cell and pore-scale network model.

For the research reported in this paper, two series of drainage simulations are performed: quasi-static and dynamic simulations. The corresponding boundary conditions are sketched in figure 6. In the first series of simulations, a small fluid pressure is imposed on the nonwetting fluid reservoir. Fluid interfaces are tracked until they reached equilibrium positions. Then, inflow boundary pressure is increased with an increment and the interfaces are moved to new equilibrium positions. This procedure is repeated until the boundary pressure is equal to the boundary pressure used in the dynamic drainage experiment. Only at the equilibrium points, volume averaged P^n, P^w and S^w are determined. At equilibrium, phase pressures P^n and P^w are equal to their corresponding reservoir pressures. From this point onward, this series is called the *quasi-static drainage experiment*.

In the second series, a large fluid pressure is imposed on the nonwetting fluid reservoir at time $t = 0$. Fluid interfaces are tracked in time until equilibrium between both fluid phases is achieved. At each moment in time, volume averaged P^n, P^w and S^w are determined. From this point onward, this is referred to as the *dynamic drainage experiment*.

In this research, only primary drainage is considered. Pore bodies and throats follow a truncated log-normal distribution. Initially, the network is fully saturated with wetting fluid. Fluid properties are partly based on those of PCE (perchloroethylene) and water and are given in table 1.

Table 1. Fluids' parameters

Contact angle (^{o})	0.0
Interfacial tension ($kg\ s^{-2}$)	0.0475
Wetting fluid viscosity ($kg\ m^{-1}s^{-2}$)	0.001
Nonwetting fluid viscosity ($kg\ m^{-1}s^{-1}$)	0.010

The initial pressure is 0 kPa. Throughout all simulations, pressure at the wetting fluid reservoir is kept constant at 0 Pa. Pressure at the nonwetting fluid reservoir is different for the quasi-static and dynamic drainage experiments. In quasi-static experiments, pressure at the nonwetting fluid boundary is initially 3 kPa. This value increases with increments of 50 Pa every time equilibrium between the fluid phases has been achieved, until a given end value is reached.

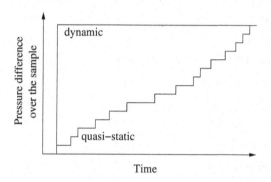

Figure 6. Fluid reservoir pressure difference in time for a quasi-static and a dynamic experiment.

3.2. DETERMINATION OF DAMPING COEFFICIENT τ

Results of simulations of quasi-static drainage experiments and various dynamic experiments, together with equation (1), can be used to determine τ as a function of saturation. Values of dynamic and quasi-static capillary pressure and saturation rate of change are obtained from our simulations at any given saturation. These are then substituted in equation (1) to obtain damping coefficient τ at that saturation. Following this procedure, there is a curve for τ for each of the chosen saturations.

The pore-scale network consists of 10×10×82 pore bodies. Geometrical properties are given in table 2. Pressures in the nonwetting reservoir are 10 kPa, 15 kPa, 20 kPa and 25 kPa for the different dynamic drainage experiments. For the quasi-static drainage, end value of the pressure in the nonwetting fluid reservoir is set to 25 kPa.

3.3. ROLE OF AVERAGING DOMAIN SIZE

To investigate the role of the averaging domain size on capillary pressure-saturation curves, three networks are constructed with identical pore and throat distribution, but with different sizes: 10×10×100, 10×10×82 and 10×10×52. Geometrical properties are given in table 2.

First, quasi-static drainage experiments are performed. Initial value of the pressure in the nonwetting fluid reservoir is 3 kPa, end value is 10 kPa. Then, the same experiment is performed under dynamic conditions. That is, fluid pressure in the nonwetting fluid reservoir is kept constant at 10 kPa.

Averaging is performed over middle layers of the network. The amount of layers to be skipped from the boundary varies from 0 to 25. Average wetting fluid saturation and average capillary pressure are determined from the averaging domain.

Table 2. Lattice network parameters

Lattice dimensions	10×10×82	10×10×100	10×10×52
Coordination number	6	6	6
Lattice spacing (10^{-4} m)	3.0	3.0	3.0
Min pore body radius (10^{-4} m)	0.16	0.16	0.16
Max pore body radius (10^{-4} m)	1.5	1.5	1.5
Mean pore body radius (10^{-4} m)	0.86	0.86	0.86
Min pore throat radius (10^{-4} m)	0.02	0.02	0.02
Max pore throat radius (10^{-4} m)	0.3	0.3	0.3
Mean pore throat radius (10^{-4} m)	0.16	0.16	0.16
Standard deviation of distribution	0.25	0.25	0.25

4. Results

4.1. DETERMINATION OF DAMPING COEFFICIENT τ

As mentioned before, the 10×10×82 network is used to determine τ. The capillary pressures and saturations in the dynamic and quasi-static simulations were averaged over the whole network. Based on the results shown in figure 7, it is decided to determine τ for saturations 0.4 through 0.9. Since irreducable saturation for the quasi-static drainage experiment is 0.33, the difference between dynamic and quasi-static can only be determined for larger saturation values. That the irreducable saturation of the quasi-static capillary pressure curve is larger than that of the dynamic curves is also visible in figure 8, where saturation in time is shown for different dynamic boundary conditions. The explanation is that the lower boundary conditions during quasi-static drainage give more possibilities for bypassing. This increases the chance of wetting fluid becoming trapped. The trapped fluid contributes to the irreducable saturation.

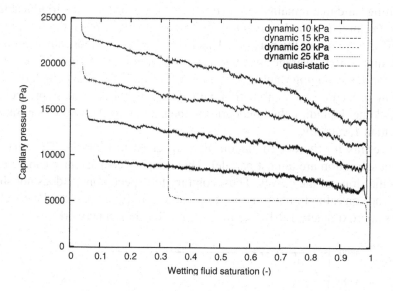

Figure 7. Dynamic and quasi-static capillary pressure curves computed from the 10×10×82 network using all layers in the averaging. These curves are used in the determation of τ.

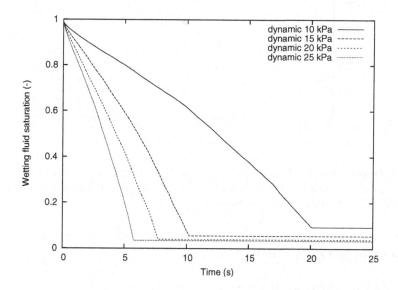

Figure 8. Time-saturation curves for the same network as in figure 7. These curves provide information on saturation rate of change as a function of saturation.

The kink in the curves shown in figure 8 is due to the fact that under dynamic conditions, the nonwetting fluid front passes like a piston through the network, as the nonwetting fluid is much more viscous than the wetting fluid (see table 1). Since it is not possible for nonwetting fluid to enter the wetting fluid reservoir, drainage stops when the nonwetting fluid front reaches the wetting fluid reservoir. All wetting fluid that remains behind the front then adds up to the irreducable saturation. For higher boundary conditons, the piston will displace the wetting fluid more effectively out of the network. This leads to a lower irreducable saturation.

The high initial average dynamic capillary pressures in figure 7 are not an artefact of the averaging routine. It is an effect of including the boundaries in the averaging procedure, in combination with averaging for all time steps, including the initial one. Including the boundaries in the averaging also causes saturation to be less than 1, initially.

The results for the determination of τ are given in figure 9. The curves for different saturation are almost parallel. The damping coefficient τ can be evaluated as the slope of the curves. Thus, contrary to expectations, τ does not show a dependence on saturation. From the curves shown in figure 9, it can be concluded that τ is approximately $1.2\cdot10^5$ kg m^{-1} s^{-1} for the chosen network.

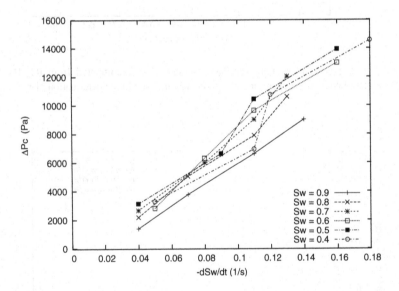

Figure 9. Determination of damping coefficient τ. Plotted are differences between dynamic and quasi-static capillary pressure curves (ΔPc) versus $-\frac{dS^w}{dt}$ for saturations 0.4 through 0.9. τ is defined as the slope of these curves, and is approximately $1.2\cdot10^5$ kg m^{-1} s^{-1}.

4.2. ROLE OF AVERAGING DOMAIN SIZE

Quasi-static drainage experiments
The results of the 10×10×100 quasi-static drainage experiment are given in figure 10. The averaging is performed over the middle 50 and 70 layers, and over all

100 layers. When all layers are included in the averaging domain, saturation is less than 1, since the nonwetting reservoir is taken into account as well. Similar results that were obtained with a 10×10×82 and a 10×10×52 network are shown in figures 11 and 12. For former simulation, averaging is performed over the middle 60 and 70 layers, and over all 82 layers. For latter, averaging is performed over the middle 30 and 40 layers, and over all 52 layers. All three figures 10–12 show that the average quasi-static capillary pressure curves are independent of the size of the averaging domain.

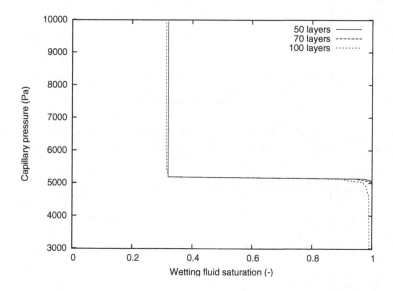

Figure 10. Numerical results of the 10×10×100 network. Shown are capillary pressure-saturation curves from the quasi-static experiments for different averaging domain sizes.

Dynamic drainage experiments

To keep the output readable, a moving-median filter has been applied to the results of the dynamic drainage experiments described in this section. Filter length is 200 data points. This filtering has no effect on the results and conclusions of our simulations.

Figure 13 shows dynamic capillary pressure-saturation curves for a 10×10×100 network. Averaging is performed over the middle 50 and 70 layers, and over all 100 layers. What is evident, is that the dynamic capillary pressure increases for increasing averaging domain size. This is due to the the chosen fluid properties in combination with the Dirichlet boundary conditions. Most of the pressure drop over the network will be in the nonwetting fluid as it is much more viscous. An increase of nonwetting boundary pressure will therefore affect the average nonwetting fluid pressure more than the average wetting fluid pressure. Thus, capillary pressure will increase for increasing boundary conditions.

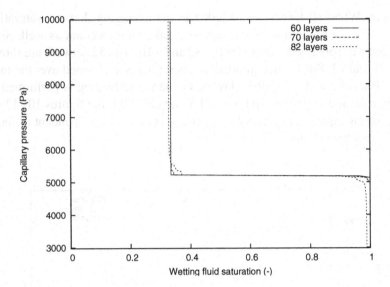

Figure 11. Numerical results of the 10×10×82 network. Shown are capillary pressure-saturation curves from the quasi-static experiments for different averaging domain sizes.

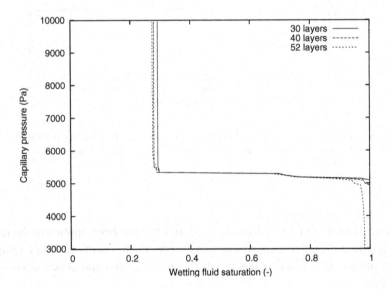

Figure 12. Numerical results of the 10×10×52 network. Shown are capillary pressure-saturation curves from the quasi-static experiments for different averaging domain sizes.

The irreducable saturation of the quasi-static capillary pressure curve in figure 10 is larger than the one of the dynamic curves in figure 13. This is a result of trapping of the wetting fluid. The explanation is the same as in section 4.1.

In figure 14, saturation versus time is shown for different averaging domain sizes in a 10×10×100 network. It is clear that $-\frac{dS^w}{dt}$ is larger for the smaller

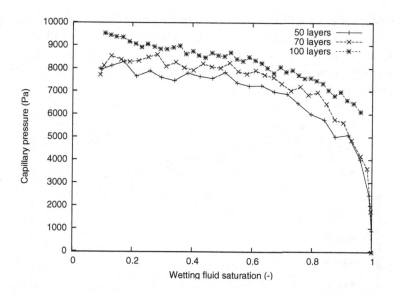

Figure 13. Numerical results of the 10×10×100 network. Shown are capillary pressure-saturation curves from the dynamic experiments for different averaging domain sizes. Averaging is performed over the center 50 (bottom curve), 70 and all 100 layers (top curve).

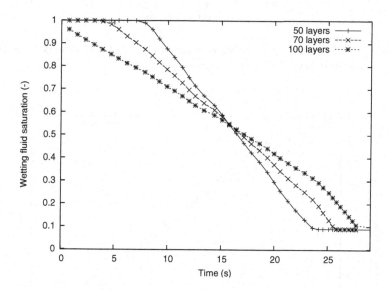

Figure 14. Numerical results of the 10×10×100 network. Shown are saturation versus time from the dynamic experiments for different averaging domain sizes. Averaging is performed over the same domains as in figure 13.

averaging domains. Under the imposed dynamic conditions, the nonwetting fluid moves as as stable front through the network, as the nonwetting fluid is much more viscous. A small domain is then drained faster than a large domain.

A similar set of experiments was performed with a 10×10×82 network and a 10×10×52 network, see figures 15–18. The same observations as with the 10×10×100 network can be made. Again, upscaled dynamic capillary pressure is higher for a larger averaging domain. Here also, the dynamic irreducable saturations are lower here than the quasi-static one.

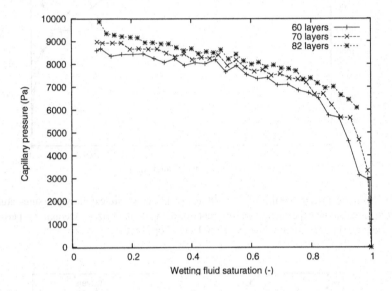

Figure 15. Dynamic capillary pressure-saturation curves of a 10×10×82 network. Averaging is performed over 60 (bottom curve), 70 and 82 layers (top curve).

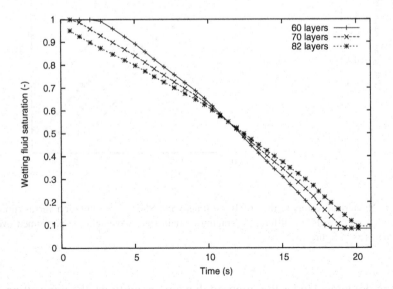

Figure 16. Time-saturation curves for the same network and averaging domains as in figure 15.

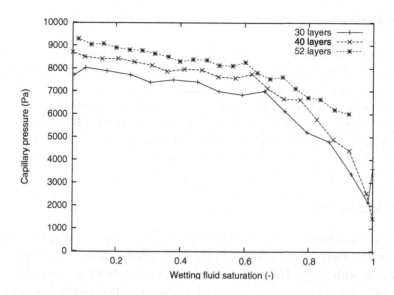

Figure 17. Dynamic capillary pressure-saturation curves of a 10×10×52 network. Averaging is performed over 30 (bottom curve), 40 and 52 layers (top curve).

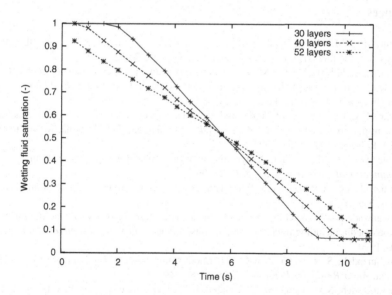

Figure 18. Time-saturation curves for the same network and averaging domains as in figure 17.

Without computing τ explicitly, one can see from figures 13–18 that the value of τ increases with increasing averaging domain size. Firstly, the difference between the dynamic and quasi-static average capillary pressure curve increases. Secondly, the rate of change of saturation $\frac{dS^w}{dt}$ decreases. Both changes contribute to an increase of damping coefficient τ.

5. Conclusions

In this paper, a dynamic pore-scale network model is presented and used to investigate an extended relationship between capillary pressure and saturation. In this definition, a dynamic correction term appears, with coefficient τ. To investigate this coefficient, quasi-static and dynamic experiments are simulated.

For the chosen network, the value of damping coefficient τ is determined to be $1.2 \cdot 10^5$ kg m^{-1} s^{-1}. Preliminary results indicate that the value of τ increases with increasing averaging domain in a fixed-size network.

Acknowledgements

This research has been carried out in the framework of project "Upscaling", financed by the Dutch Science Foundation (NWO), under grant number 809.62.010. Discussions with Helge Dahle (University of Bergen, Norway) and Mike Celia (Princeton University, USA) are greatly appreciated. Denis Demidov (Kazan State University, Russia) is thanked for his help with on the trapping algorithm.

References

1. Blunt, M. & M.J. King, Relative permeabilities from two- and three-dimensional pore-scale network modelling, *Transport in Porous Media*, **6**, pp. 407-432, 1991
2. Celia, M.A., H.K. Dahle & S.M. Hassanizadeh, Dynamic pore-scale network models for two-phase flow in porous media, in: L. Bentley et al. (Eds.), Computational Methods in Water Resources, Vol. 1, A.A. Balkema, pp. 217-223, 2000
3. Celia, M.A., P.C. Reeves & L.A. Ferrand, Recent advances in pore scale models for multiphase flow in porous media, *Reviews of Geophysics*, Supplement, U.S. National Report to IUGG, 1991-1994, pp. 1049-1057, 1995
4. Dahle, H.K. & M.A. Celia, A dynamic network model for two- phase immiscible flow, *Computational Geosciences*, **3**, pp. 1-22, 1999
5. Dullien, F.A.L., Porous Media: Fluid Transport and Pore Structure, 2nd edition, Academic Press, New York, 1992
6. Hassanizadeh, S.M. & Gray, W.G., Mechanics and thermodynamics of multiphase flow in porous media including interphase boundaries, *Advances in Water Resources*, **13(4)**, pp. 169-186, 1990
7. Hassanizadeh, S.M. & Gray, W.G., Thermodynamic basis of capillary pressure in porous media, *Water Resources Research*, **29**, pp. 3389-3405, 1993
8. Hassanizadeh, S.M., Celia, M.A. & Dahle, H.K., Dynamic effects in the capillary pressure-saturation relationship and their impact on unsaturated flow, *Vadose Zone Journal*, **1**, pp. 38-57, 2002
9. Mogensen, K. & E.H. Stenby, A dynamic two-phase pore-scale model of imbibition, *Transport in Porous Media*, **32**, pp. 299-327, 1998
10. Smiles, D.E., Vachaud, G. and Vauclin, M., A test of the uniqueness of the soil moisture characteristic during transient, non-hysteretic flow of water in a rigid soil, *Soil Sci. Soc. Amer. Proc.*, **35**, pp. 535-539, 1971
11. Stauffer, F., Time dependence of the relations between capillary pressure, water content and conductivity during drainage of porous media, *Proceedings International IAHR Symposium on Scale Effects in Porous Media, Thessaloniki, Greece, August 29 - Sept. 1*, 3.35-3.52, 1978

12. Topp, G.C., Klute, A., and Peters, D.B., Comparison of water content-pressure head data obtained by equilibrium, steady-state and unsteady-state methods, *Soil Sci. Soc. Amer. Proc.*, **31**, pp. 312-314, 1967
13. Washburn, E.W., The dynamics of capillary flow, *Physical Review*, **17(3)**, pp. 273-283, 1921

Authors' Vitae

T. Gielen
Twan Gielen is PhD student at Faculty of Civil Engineering and Geosciences of Delft University of Technology, The Netherlands. His PhD research is on upscaling multiphase transport properties from pore to core. He holds an MSc degree in Geophysics from Utrecht University, The Netherlands.

S.M. Hassanizadeh
Majid Hassanizadeh is recently appointed (starting January 1, 2004) as Professor of Hydrogeology at the Faculty of Geosciences of Utrecht University, The Netherlands. He graduated with honors from the Department of Civil Engineering of Pahlavi University in Shiraz, Iran in 1975. Earned M.E. in 1976 and Ph.D. in 1979 at the Department of Civil Engineering of Princeton University. After he worked as Assistant Professor in Iran, he became senior researcher with the National Institute of Public Health and Environment, RIVM (Bilthoven, The Netherlands, 1984-1995), Associate Professor (1995-2001) and later Professor (2001-2003) with the Faculty of Civil Engineering and Geosciences, Delft University of Technology. Majid has more than 100 publications in journals, books, conference proceedings, or as technical reports. He was editor of Advances in Water Resources from 1991 till 2001 and is now on the editorial board of Transport in Porous Media, Vadoze Zone Journal, and Advances in Water Resources. He is member of American Geophysical Union, Soil Science Society of America, European Geophysical Union, International Association of Hydrological Sciences, International Association for Environmental Hydrology, The Netherlands Hydrological Society. Elected Fellow of American Geophysical Union in 2002. He is also affiliated with Netherlands Institute of Applied Geoscience, at the Department of Geo-Energy.

A. Leijnse
Toon Leijnse holds the parttime chair Groundwater Quality in the group Soil Physics, Agrohydrology and Groundwater Management in the sub-department Water Resources of Wageningen University since 1994. He graduated from Delft University of Technology, Faculty of Physical Engineering in 1970, and received a Ph.D. from the University of Notre Dame (USA), Dept. of Civil Engineering and Geological Sciences in 1992. He is senior research scientist at the Dutch National Institute for Public Health and the Environment, and since 2000, he is associated as a senior research scientist with the Netherlands Institute of Applied Geoscience TNO. Major areas of research are groundwater flow and reactive

transport, density-dependent flow and transport in porous media and parameter estimation and uncertainty analysis. He is a member of the Americal Geophysical Union, the International Association of Hydrological Sciences and the International Association of Hydraulic Research, where he is a member of the International Groundwater Committee. He (co-)authored some 100 publications among which a book on "Mathematical Tools for Changing Scales in Physical Systems".

H.F. Nordhaug

Hans Fredrik Nordhaug is researcher at Centre for Integrated Petroleum Research (CIPR), University of Bergen, Norway. He holds an MSc degree and a PhD degree in Applied Mathematics from Bergen University. For his PhD degree, he has done extensive work on pore-scale network models for testing multiphase theories including interfacial area and dynamic capillary pressure.

UPSCALING OF TWO-PHASE FLOW PROCESSES IN POROUS MEDIA

HARTMUT EICHEL, RAINER HELMIG, INSA NEUWEILER, and OLAF A. CIRPKA
Universität Stuttgart, Institut für Wasserbau,
Pfaffenwaldring 61, 70569 Stuttgart, Deutschland

Abstract. Intrinsic heterogeneities influence the multi-phase flow behavior of a dense non-aqueous phase liquids (DNAPL) infiltrating into a natural soil. Typically, we cannot resolve the scale of these heterogeneities so that upscaling techniques are required. The choice of the appropriate upscaling method depends on the averaging scale, since the relative importance of capillary and gravity forces change with scale. We present an easy and quick upscaling approach for cases in which the flow on the length-scale of heterogeneities is dominated by capillary forces.

The approach is based on a percolation model and a single-phase flow-averaging method. We apply the upscaling approach to experimental data of a DNAPL infiltration into a sandbox with artificial sand lenses. The anisotropy of the structure results in anisotropic flow which is amplified by the non-linear behavior of multi-phase flow. The residual saturation depends on the direction of flow, and the anisotropy ratio of the effective permeability is a function of the DNAPL saturation. Furthermore, it appears necessary to regard the relative permeability–saturation relationship as a tensor property rather than a scalar. The overall flow behavior simulated by the upscaled model agrees well with simulations accounting for the distinct lenses and the experimental data.

1. Motivation

Multi-phase flow and transport processes in porous media play an important role in the remediation of non-aqueous phase liquids (NAPL) in the subsurface. These flow processes are affected by heterogeneities on all scales. Spatial variability ranges from single pores to geological structures, thereby spanning length scales from μm to km (see Figure 1).

Although the term *pore scale* is unambiguous, all other terms describing scales like micro or macro scale are not necessarily consistently used. For example, the typical length scales considered in petroleum engineering differ significantly from those in environmental engineering. While oil fields extend over hundreds of meters to kilometers, the typical scales for environmental problems range from meters to tens of meters (see Figure 2).

Different forces are likely to dominate on different scales. While on smaller scales capillary effects are more pronounced, the gravity effects and the viscous effects become more important at larger scales. For an environmental engineer, therefore, both the capillary and the gravity forces may be important. The dom-

D.B. Das and S.M. Hassanizadeh (eds.) Upscaling Multiphase Flow in Porous Media, 237–257
© 2005 Springer. Printed in the Netherlands

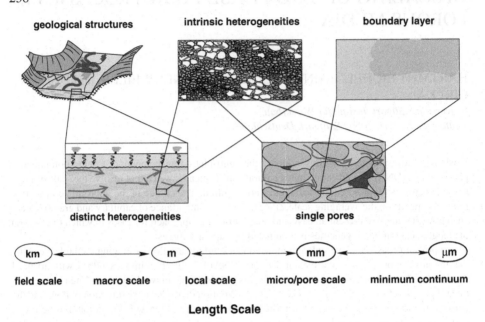

Figure 1. Scales in subsurface hydrology (after Kobus, de Haar, 1995)

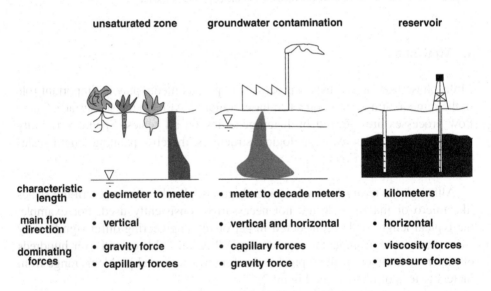

Figure 2. Different scales for different applications.

inating forces need to be determined prior to choosing a particular upscaling method.

In our example, we analyze an experiment carried out by (Allan et al., 1998; Kobus et al., 2000) at the research facility for subsurface remediation, VEGAS, at the University of Stuttgart. In the experiment, a dense non-aqueous phase liquid

(DNAPL) was infiltrated into a small sandbox. Figure 3 shows the distribution of sand types and the dimensions of the domain. The typical length scale is on the order of centimeters to decimeters, where gravity forces as well as capillary forces have to be considered.

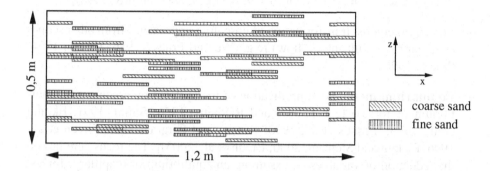

Figure 3. Distribution of the sand types in the experiment.

It is well known that small-scale structures, such as small-scale sand lenses, can influence multi-phase flow and transport significantly. Many laboratory experiments have shown that capillary forces have a large impact on the two-phase flow behaviour in porous media on almost all scales (Jakobs et al., 2003; Illangasekare et al., 1995) .

This paper focuses on heterogeneities within larger-scale structures as seen in Figure 3. As is commonly known, the NAPL cannot penetrate a region of finer material as long as the capillary pressure has not yet exceeded the entry pressure. Thus, small-scale layering can lead to significant lateral spreading of the NAPL.

In spite of increasing computational power, the simulation of multi-phase flow in porous media is still restricted to comparably coarse grids, prohibiting the resolution of small-scale features. In the simulation of multi-phase flow on larger scales, it is therefore necessary to parameterize the effects of small-scale heterogeneities on the large-scale flow behavior. A variety of techniques have been developed and applied to transfer the information from the process scale to the simulation scale. These techniques are commonly referred to as upscaling. The upscaling techniques may be classified into the following categories:

- A-posteriori methods (effective parameters are derived from the analysis of highly resolved computation or measurement) [e.g. (Christie, 1996; Pickup and Sorbie, 1996; Dale et al., 1997; Chang and Mohanty, 1997)]

- Stochastic methods (determination of the effective parameters through assumptions of the statistical distribution of the heterogeneities and a stochastical averaging of the equations) [e.g (Desbarats, 1995; Yeh et al., 1985; Man-

toglou and Gelhar, 1987; Chang et al., 1995; Neuweiler et al., 2003; Efendiev
and Durlofsky, 2003)].

– Analytical methods (computation of the effective parameters for simple configurations, volume averaging) [e.g. (Quintard and Whitaker, 1988; Ahmadi
and Quintard, 1996; Saez et al., 1989; Bourgeat and Panfilov, 1998)]

– Analogy methods (transformation of upscaling-approaches from other scopes
of research to multi-phase flow) [e.g. (Wu and Pruess, 1986; Pruess, 1994;
Pruess, 1996)]

– **Equilibrium methods** (simplification of the systems by assuming an equilibrium of forces) [e.g. (Corey and Rathjens, 1956; Smith, 1991; Ekrann
et al., 1996; Yortsos et al., 1993; Kueper and Girgrah, 1994; Green et al.,
1996; Pickup and Stephen, 2000; Braun et al., 2005)]. The focus here lies on
the reduction of variables by assuming an equilibrium of (capillary) forces.
This assumption allows for the application of a percolation model which, together with an appropriate averaging method, results in effective constitutive
relationships for the macroscale.

The purpose of our upscaling approach is to compromise between two goals. First,
we want to develop a relatively easy method, which is not restricted to a certain set
of boundary and flow conditions and therefore applicable to different scenarios.
Second, we want to reproduce the most important physical effects.

This paper is organized as follows. In Section 2, we present the experimental
model set-up. We evaluate the validity of the capillary-equilibrium assumption by
a dimensional analysis in Section 3. In Section 4, we introduce our upscaling approach. We compare the results of different models and the experiment in Section
5. In Section 6, finally, we draw conclusions and give an outlook to future studies.

2. Physical Model Setup

The physical experiment that we use as a reference was carried out by (Allan et
al., 1998; Kobus et al., 2000).

2.1. PROPERTIES OF THE POROUS MEDIUM

A DNAPL is infiltrated into a water-saturated sandbox of dimensions
(L×W×H) 1.2 m × 0.08 m × 0.5 m. The porous medium comprises three different
sands, a fine, a medium, and a coarse one. The medium sand is the background
material, while the other two are incorporated as lenses with a width of 0.2 m and
a height of 0.01 m, see Figure 3. The properties of the sands are listed in Table 1.
The medium sand occupies 80% of the domain whereas the fine and coarse sand
take each 10%. The lenses are randomly distributed.

Table 1. Properties of the sands used in the experiment.

sand	fine	medium	coarse
permeability k [m^2]	$6.38 \cdot 10^{-11}$	$1.22 \cdot 10^{-10}$	$2.55 \cdot 10^{-10}$
entry pressure P_d [Pa]	882.9	539.55	353.16
form factor λ [-]	3.0	3.0	3.2
residual saturation wetting phase S_{wr} [-]	0.06	0.06	0.06
residual saturation non-wetting phase S_{nr} [-]	0.10	0.15	0.10
porosity ϕ [-]	0.38	0.38	0.38
volume w_i [%]	10	80	10

2.2. BOUNDARY AND INITIAL CONDITIONS

Initially, the entire domain is fully water-saturated, and there is no flow. This results in a hydrostatic pressure distribution. Over the entire course of the experiment, the left and the right faces of the domain are connected to a water reservoir ensuring constant pressure conditions at the boundaries. Water that is replaced by the infiltrating DNAPL can leave the system over these boundaries. The bottom and top boundaries are non-permeable for both liquids, except for a small stretch of two centimeters in the top, where the DNAPL infiltrates with a rate of $0.4833 \cdot 10^{-6}$ m^3/s for 2970 seconds. The initial and the boundary conditions are depicted in Figure 4.

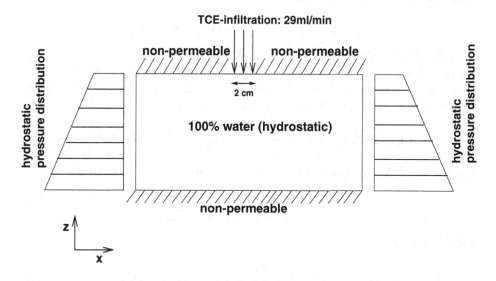

Figure 4. Initial and boundary conditions.

3. Mathematical Model and Dimensional Analysis

Our upscaling approach is based on capillary equilibrium. This means, that capillary forces are dominant on the small scale. In the following, we test this assumption by a dimensional analysis.

The system of the two-phase flow equations expresses the conservation of mass and generalized Darcys Law o f both fluids. We assume a rigid solid phase and incompressible fluids. The following equations describe the flow on the local scale.

$$\phi(\vec{x})\frac{\partial S_w}{\partial t} + \vec{\nabla} \cdot \left(-\frac{k_{r,w}(S_w, \vec{x})}{\mu_w}k(\vec{x})\left(\vec{\nabla}(P_w) + \varrho_w\vec{g}\right)\right) = q_w \qquad (1)$$

$$\phi(\vec{x})\frac{\partial S_n}{\partial t} + \vec{\nabla} \cdot \left(-\frac{k_{r,n}(S_n, \vec{x})}{\mu_n}k(\vec{x})\left(\vec{\nabla}(P_n) + \varrho_n\vec{g}\right)\right) = q_n \qquad (2)$$

where ϕ, ϱ, S, μ, g, P, and q are the porosity, the density, the saturation, the viscosity, the acceleration constant due to gravity, the phase pressure and the source/sink term, respectively. k is the intrinsic permeability, k_r is the relative permeability of the respective phase, and $k_{eff} = k_r k$ is the effective permeability. The subscripts n and w denote the non-wetting and the wetting phase. For a detailed derivation of this formulation see, e.g., (Marle, 1981; Helmig, 1997).

These two equations are coupled by the two following relations. First, the two saturations sum up to unity. Second, the capillary pressure, defined as the difference between the pressure of the non-wetting and the wetting fluids, is a unique function of the saturation.

$$S_n + S_w = 1, \qquad P_c(S_n) = P_n - P_w. \qquad (3)$$

In the dimensional analysis, we assume that the source/sink terms, q_w and q_n, are zero. We sum up the two equations and introduce the total Darcy velocity $\vec{u}_{total} = \vec{u}_n + \vec{u}_w$. In that way, we eliminate the phase pressures and the phase Darcy velocities from the equation. The two-phase continuity equation reads:

$$\phi\partial_t S_n + \vec{u}_{total} \cdot \vec{\nabla}f(S_n) + \vec{\nabla} \cdot \left(\frac{gk\Delta\rho}{\mu_n}\Lambda(S_n)\vec{e}_z\right) - \vec{\nabla} \cdot \left(\frac{k}{\mu_n}\Lambda(S_n)\vec{\nabla}P_c\right) = 0. \qquad (4)$$

The fractional flow function $f(S_n)$ of the non-wetting fluid is defined as:

$$f(S_n) = \frac{k_{r,n}(S_n)}{k_{r,n}(S_n) + \frac{\mu_n}{\mu_w}k_{r,w}(S_n)}, \qquad (5)$$

and $\Lambda(S)$ stands for

$$\Lambda(S_n) = \frac{\frac{\mu_n}{\mu_w}k_{r,n}(S_n)k_{r,w}(S_n)}{k_{r,n}(S_n) + \frac{\mu_n}{\mu_w}k_{r,w}(S_n)} = \frac{\mu_n}{\mu_w}k_{r,w}f(S_n). \qquad (6)$$

The density difference is defined as $\Delta\rho = \rho_n - \rho_w$.

As a first approximation we assume that we can neglect horizontal total flow velocities due to pooling and have therefore only capillary forces acting in horizontal direction.

We introduce typical scales for time, length, and capillary pressure. The time is scaled with gravity. Typical length scales are the dimensions of the domain, X and Z, and the capillary pressure is scaled by the entry pressure P_d,

$$t^\star = \frac{t\,k\,\Delta\rho\,g}{Z\,\mu_n}, \qquad z^\star = z/Z, \qquad x^\star = x/X, \qquad P_c^\star = P_c/P_d \qquad (7)$$

in which the star denotes dimensionless variables. Thus, we obtain the following dimensionless form of equation 4:

$$\partial_{t^\star} S_n + \mathbf{Gr}^{-1}\partial_{z^\star} f(S_n) + \partial_{z^\star}\Lambda(S_n) - \mathbf{Bo_z}^{-1}\partial_{z^\star}\Lambda(S_n)\partial_{z^\star} P_c^\star(S_n) \qquad (8)$$
$$-\mathbf{Bo_x}^{-1}\partial_{x^\star}\Lambda(S_n)\partial_{x^\star} P_c^\star(S_n) = 0$$

with the inverse gravity number $\mathbf{Gr}^{-1}$ and the inverse Bond numbers $\mathbf{Bo}^{-1}$ in the x direction and in the z direction.

$$\mathbf{Gr}^{-1} := \frac{\text{viscous forces}}{\text{gravity forces}} = \frac{\mu \cdot u}{\Delta\rho \cdot g \cdot k} \qquad (9)$$

$$\mathbf{Bo_z}^{-1} := \frac{\text{capillary forces}}{\text{gravity forces}} = \frac{P_d}{\Delta\rho \cdot g \cdot Z} \qquad (10)$$

$$\mathbf{Bo_x}^{-1} := \frac{\text{capillary forces}}{\text{gravity forces}} = \frac{P_d \cdot Z}{\Delta\rho \cdot g \cdot X^2} \qquad (11)$$

The capillary effects are accounted for by the Bond numbers. As an alternative, one may use the capillary number $\mathbf{Ca}$, which is related to the Bond number by:

$$\mathbf{Ca} = \frac{\text{capillary forces}}{\text{viscous forces}} = \frac{\mathbf{Gr}}{\mathbf{Bo}}. \qquad (12)$$

We evaluate these quantities on the large (domain) scale. Considering the typical parameter values for the background sand material $k = 1.22 \cdot 10^{-10} \ m^2$, $P_d = 540 \ Pa$, the length scales of the domain $Z = 0.5 \ m$, $X = 1.2 \ m$, and the liquid properties $\Delta\rho = 460 \ kg/m^3$, $\mu_n = 5.7 \cdot 10^{-4} \ kg/(ms)$, the only quantity that we have to evaluate is the characteristic velocity u. A rough estimation is given by assuming only the vertical component. The injected volumetric flux is $Q = 4.8 \cdot 10^{-7} \ m^3/s$, the width of the inlet is 2 cm, and the box is 8 cm thick. This yields a maximum vertical darcy velocity of $u_{\text{total}} = 3 \cdot 10^{-4} \ m/s$. We assume that the velocity of the wetting phase is negligible. If we insert these values into equations 9 – 11, we get the following values of the three characteristic dimensionless numbers.

$$\mathbf{Gr}^{-1} = 0.31, \quad \mathbf{Bo_z}^{-1} = 0.24, \quad \mathbf{Bo_x}^{-1} = 0.04. \tag{13}$$

For the detailed flow process we have to consider the small length scale, given by the dimension of the inclusions. If we introduce the ratio between the length scales of the inclusions and the domain (Duijn et al., 2002),

$$\epsilon_{vert.} = \frac{\Delta z}{\Delta Z} = 0.02, \qquad \epsilon_{horiz.} = \frac{\Delta x}{\Delta X} = 0.17, \tag{14}$$

the spatial derivatives on the small and large scales can be separated. Having chosen the base system to be the large (domain) scale, the spatial derivatives on the small scale are multiplied by a factor of $1/\epsilon$. Since the expressions accounting for "diffusive" processes have second-order spatial derivatives, they are scaled by $1/\epsilon^2$ on the small scale, whereas the "advective" processes are scaled by $1/\epsilon$ on the small scale. In this way, the capillary processes are "magnified" on the small scale, and their impact is higher than on the large scale. If the respective inverse Bond numbers are small (the same order as ϵ), the "magnification" of the capillary processes on the small scale cancels out and advective and diffusive processes contribute on the small scale to the same extent.

If the capillary number is of order 1 and large compared to ϵ, the diffusive processes on the small scale are weighted by $1/\epsilon$ compared to the advective processes. In this case, the small scale is dominated by capillary forces, and the viscous and gravity forces on this scale can be neglected.

In order to meet the criterion for capillary dominance, a clear separation of scales must be given:

$$\epsilon \ll \mathbf{Bo}^{-1} \ll 1/\epsilon \quad \epsilon \ll \mathbf{Ca}^{-1} \ll 1/\epsilon. \tag{15}$$

In our case, the separation criterion is met in the vertical direction, $\epsilon = 0.02 < \mathbf{Bo_z}^{-1} = 0.24 \ll 1/\epsilon = 50$. As the inverse gravity number is between 0.1 and 1, the criterion (15) is also met for the inverse capillary number.

Although the criterion for capillary equilibrium is met in the experiment considered here, the following points should be considered:

– The analysis holds only when the inclusions are placed in distances in the same range as the length scale of the inclusion. Otherwise we get the average distance of the inclusions as an additional intermediate scale. Also, the contrast of the parameter properties, such as permeability and capillary entry pressure has to be large compared to ϵ and small compared to $1/\epsilon$. Obviously the difference of the function $\Lambda(S_n)$ in the different materials should also not be large, in order to keep the scales separated.

- We could also use the small scale as the base system and scale all numbers accordingly with the small length scales. By this, we would obtain identical results.

- The actual experimental setup is more complex. The influx is not placed over the whole width of the tank. The estimation for the resulting vertical and horizontal total flow velocities is therefore not trivial.

4. Upscaling Method

In this section, we describe the different underlying assumptions and the steps comprising the proposed upscaling approach, used to derive effective parameters on the macroscale for the simulation of two–phase flow processes. This is done by a percolation model and a small-scale flow-averaging method.

4.1. ASSUMPTIONS

As outlined above, we assume that capillary effects dominate the processes on the small scale. Changes of variables on the large scale are very slow compared to changes of variables on the small scale. From the perspective of the large scale, this implies that the small-scale reaction on a change of large-scale variables is quasi instantaneously. Thus, we can neglect the dynamics on the small scale and assume, on that scale, that the system is in capillary equilibrium. We make use of that property in a percolation model for the small-scale features. Here, we assume that, given a large-scale capillary pressure, the non-wetting phase enters instantaneously all cells of the small-scale model in which the entry pressure is exceeded. Therefore hysteresis does not play a role in this model. The fluid distribution in the small-scale model is given from the local $P_c - S$ relations that are represented by Brooks-Corey type functions (Brooks and Corey, 1966), with no residual saturation (S_{nr}) on the local scale. By this means, we can construct the functional relation between the capillary pressure and the large-scale saturation.

4.2. PERCOLATION MODEL

In our application, we know the exact distribution of the materials with their parameters and constitutive relationships. Applying the capillary equilibrium assumption to a distribution of local $P_c - S_w$ relationships, we can determine the saturation distribution for a given capillary pressure. We do this by applying a static site–percolation model (Stauffer, 1985). The arithmetic mean of the saturation distribution gives one point on the macroscopic capillary pressure–saturation–relationship. In Figure 5, three different capillary pressure levels and the associated macroscopic saturations are shown. The three resulting points on the macroscopic curve are shown in Figure 6.

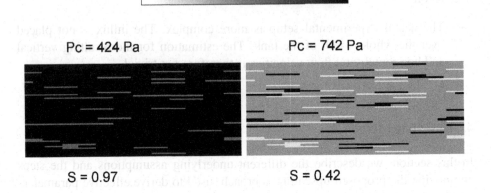

Pc = 424 Pa   Pc = 742 Pa

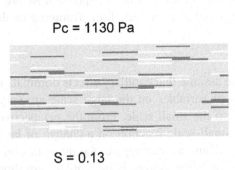

S = 0.97   S = 0.42

Pc = 1130 Pa

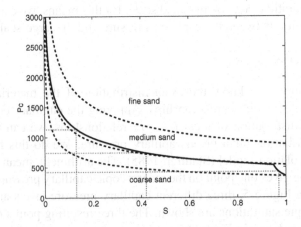

S = 0.13

Figure 5. Steps in the percolation model.

Cycling through this procedure with different capillary pressures, one can determine the complete macroscopic capillary pressure–saturation relationship.

Figure 6. Macroscopic capillary pressure – saturation relationship

In Figure 6, four different $P - c - S$ relationships are shown. The three dashed curves indicate the relationships for the three individual sands, while the solid line is the determined macroscopic P_c-S-relationship. The P_c - S relationship resembles the curve of the medium sand quite closely (Figure 6). Only at high saturations of the wetting fluid, the upscaled curve shows a dip that does not exist in the retention curve of the medium sand. At this saturation the entry pressure of the background sand is exceeded.

4.3. RENORMALIZATION

As a first upscaling approach for the relative permeabilities, we test the renormalization approach as suggested by Williams and King (Williams, 1989; King, 1996). For a quadratic domain the effective horizontal conductivity $\overline{k_h}$ can be computed by a finite difference method. The indices are shown in Figure 7.

$$\overline{k_h} = \frac{2 \cdot (k_1 + k_2) \cdot (k_3 + k_4) \cdot (k_{12}^h + k_{34}^h)}{3 \cdot (k_1 + k_3) \cdot (k_2 + k_4) + \frac{1}{2} \cdot (k_1 + k_2 + k_3 + k_4) \cdot (k_{12}^h + k_{34}^h)}, \quad (16)$$

$$\text{with} \quad k_{ij}^h = \frac{2 \cdot k_i \cdot k_j}{k_i + k_j}. \quad (17)$$

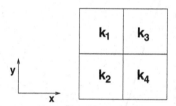

Figure 7. Indices used in the renormalization method.

For the effective vertical conductivity, the indices "2" and "3" have to be exchanged. After determining the effective conductivity of a block of four cells, one proceeds to a higher scale on which the conductivities of four blocks are averaged.

For every specific global saturation, there exists a local saturation distribution computed by the percolation model. The local $k_{eff} = k_r(S)k$ is thus known. The renormalization is performed for the effective permeability. On the highest level, the procedure results in a single effective permeability for each phase in each direction. As the relative permeability kr is defined by

$$kr_{ii} = \frac{k_{eff,ii}(S)}{k_{eff,ii}(S = 1)}, \quad \text{with } i = x, y, \quad (18)$$

the renormalization yields one point on the upscaled relative permeability - saturation relationship. Repeating this procedure with different capillary pressures, and thus different saturations, yields the two upscaled $kr_w - S_w$-relationships. This procedure is carried out for both spatial dimensions and both fluids.

It may be noteworthy that the strong anisotropy on the small scale and the harmonic weighting in the renormalization procedure yields artefacts, as can be seen in Figure 8. Here, the dashed lines indicate Brooks-Corey parameterized curves (Brooks and Corey, 1966) used as parameterizations for all materials. The solid lines represent the vertical $kr - S$-relationships computed by the renormalization method. The horizontal $kr - S$-curves, which are not shown here, are closer to the Brooks-Corey parameterizations.

The renormalization method leads to extremely high macroscopic residual saturations caused by zones of relatively low permeabilities. This may be explained by the illustrative example shown in Figure 9. In this example, a preferential, curvilinear flow path exists. The unfortunate choice of the first renormalization blocks, however, cuts the preferential flow path off. Thus, effective permeability on the highest level is strongly underestimated.

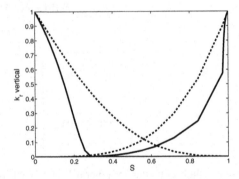

Figure 8. kr - S relationship obtained from the renormalization method.

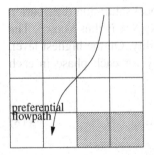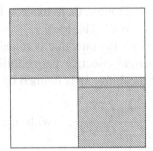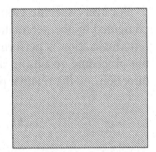

Figure 9. Renormalization techniques for anisotropic systems.

4.4. SINGLE-PHASE FLOW-AVERAGING METHOD

The upscaled relative permeability - saturation relationship can be computed by solving the pressure equation for a single phase (Dykaar and Kitanidis, 1992). Periodic boundary conditions are chosen, so that the pressure fluctuations match on opposing sides. Imposing a large-scale pressure gradient onto the system, the pressure at the inflow boundary has a higher value than that at the outflow boundary. This is accounted for by a uniform jump. The general setup of periodic cells used for upscaling is well described by (Durlofsky, 1991).

Our procedure is as follows. For a given capillary pressure, the local saturation distribution is known from the percolation model. This together with the known local kr-S relationship and the permeability distribution, yields the local effective permeability. We now solve the pressure equation for a single phase, imposing a unit pressure gradient. It is assumed that the motion of one fluid has no impact on that of the other fluid. From the pressure distribution of the single fluid we can determine its velocity field. Then, the effective permeability of the phase considered, k_{eff}, can be calculated from the mean velocity and the applied pressure gradient.

Since the relative permeability is defined as the ratio between k_{eff} (S = 1) and k_{eff} (S), the single-phase flow simulation yields a single point on the upscaled relative permeability - saturation relationship. Repeating the analysis for different capillary pressures, and thus different saturations, we construct the entire relative permeability curve. The procedure is carried out for both fluids independently.

The effective permeability value obtained is one diagonal entry in the effective permeability tensor. In order to get the second diagonal entry, another set of flow simulations is carried out, now with the pressure gradient perpendicular to the first direction. Applying periodic boundary conditions without jump along the remaining boundaries, we also determine the off-diagonal entries of the relative-permeability tensor. In the present application, however, these terms are comparably small and are thus neglected in the following analysis.

Figure 10 shows the relative permeabilities for the above explained single-phase flow averaging method, applied to the data of the sandbox. The solid lines indicate the vertical relative permeabilities, the dashed lines represent the horizontal relative permeabilities, while the dotted lines show Brooks-Corey parametrizations for the medium sand as comparison. It is clearly visible that the vertical relative permeabilities are highly reduced compared to the local Brooks-Corey curves and that the horizontal ones are slightly increased. That is, the relative permeability exhibits strong anisotropy. Also, the macroscopic residual saturations differ from the residual saturation of the Brooks-Corey curve. Both findings are in agreement with the experimental results. The lenses lead to more horizontal spreading and delay the flow in the vertical direction. In the coarse sand lenses DNAPL gets trapped, while the fine sand lenses can be bypassed. Although the

macroscopic residual saturations for flow in the vertical direction increases, they
are not as high as computed by the renormalization method.

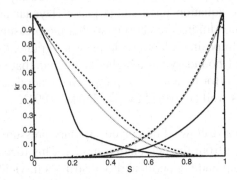

Figure 10. kr - S relationships obtained from the single-phase simulations.

Applying periodic boundary conditions in the single-phase flow simulations
implies that the domain with the small-scale features is interpreted as a unit cell of
a periodic domain, made of an infinite number of those unit cells. By construction,
the unit cell of such a system is a representative elementary volume. The periodic boundary conditions also guarantee that the resulting effective permeability
tensor is symmetric and positive-definite. When the saturation of the considered
phase becomes extremely small, however, numerical errors may cause prohibited
effective-permeability tensors.

5. Comparison of Measured and Simulated NAPL Distributions

We now compare the experimental saturation distribution (see Figure 11) with
a discrete, two-dimensional simulation (see Figure 12), in which the blocks of
different permeability are resolved explicitly. We use a boxmethod as described in
(Helmig, 1997) solving the discretized equations for water pressure and DNAPL
saturation. The grid cells are 1 cm high and 2 cm wide. The experimental results
are based on photographs taken after one hour. The exact saturation values cannot
be determined, nonetheless, the picture gives a good qualitative impression of how
far the NAPL distribution infiltrated.

The detailed simulation reproduces the experiment well with respect to the
overall NAPL distribution. The experimental data are almost binary, with NAPL
found in a few coarse-sand lenses. Here, the NAPL is entrapped by capillary
forces. The simulations predict quite well which coarse-sand blocks are occupied
by the NAPL. The simulations, however, show a higher residual saturation in the
medium-sand matrix than observed in the experiment. On the macro-scale, the
residual saturation is dominated by the entrapment in the coarse-sand lenses. In the
simulations, we can also identify some fine-sand lenses by the non-wetting phase
pooling on top of them. If we take two threshold values for the saturation, namely

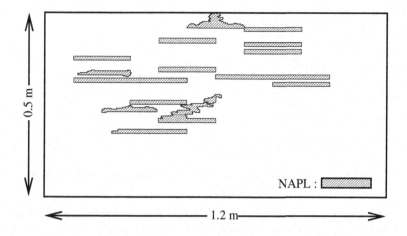

Figure 11. Experimental NAPL distribution after 1 hr (from Braun, 2000).

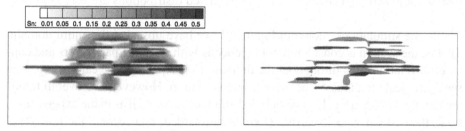

Figure 12. Discrete simulations. Left: full saturation distribution; right: two threshold values of 0.3 and 0.5.

0.3 and 0.5, we see how closely the numerical results match the experimental data.

After we have shown that the discrete simulation matches the experimental results well; we now compare these results with two simulations based on upscaled constitutive relationships. This allows us to calculate and compare first and second moments of the DNAPL body, which would have not been possible with the experimental results.

The first is a simple upscaling approach, where the permeabilities and entry pressure are just geometrically averaged to obtain the macroscopic parameters. The absolute permeability should be anisotropic due to the different correlation lengths but the influence is negligible. The relative permeabilities are approximated as Brooks-Corey parametrizations. In Figure 13, we see that the macroscopic parameters obtained by taking the geometrical average of the small scale values, cannot capture the overall flow behaviour. In this example, the downward velocity is overestimated dramatically, and the horizontal spreading is not represented.

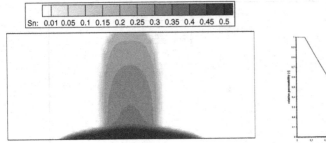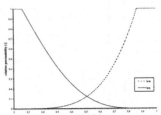

Figure 13. Saturation distribution for upscaling by taking the geometric average.

Figure 14 shows the results for two different grids using the upscaled constitutive relationships from the percolation model and the single-phase flow-approach for the relative permeabilities. The left simulation is computed on a grid which is as fine as the one used for the detailed simulations. The results shown in right subplot are obtained on a coarser grid. The predicted distributions are very similar.

For the simulations shown in Figure 14, we have upscaled the entire domain. That is, the system is considered homogeneous with identical parameters and constitutive relationships throughout the domain. Consequently, one cannot expect to see small-scale features of the saturation distribution. However, two overall trends are identical in the upscaled and detailed simulations as well as in the experiments. First, the vertical velocity of the DNAPL is retarded, and second, the horizontal spreading of the DNAPL is enhanced. The upscaled simulations reproduce those features because the vertical relative permeability curves (solid curves) seen in Figure 10 are well below the Brooks-Corey parametrizations indicated by the dotted lines, and the horizontal relative permeability curve, at least of the DNAPL, is larger. These curves reflect the effect of the lenses in the physical model that diminish downward movement of the DNAPL.

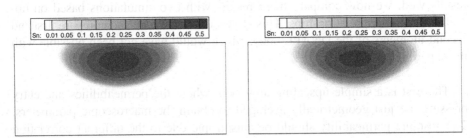

Figure 14. Saturation distribution for the upscaled anisotropic parameters – left: Fine grid simulation – right: Coarse (upscaled) grid simulation

In Figure 15, we overlay the numerical results of the discrete simulation with the proposed upscaling approach. The contour lines show the saturation distribution from the simulation with the upscaled values, while the gray and black areas indicate regions where the DNAPL exceeds a saturation of 0.3 and 0.5, respectively.

On average, the lateral spreading is matched well. Obviously, coarse-sand lenses extending beyond the region, that is reached by the DNAPL in the upscaled simulation, cannot be captured fully. The vertical migration is slightly underestimated, indicating an overestimation of macroscopic residual saturation by the upscaling approach.

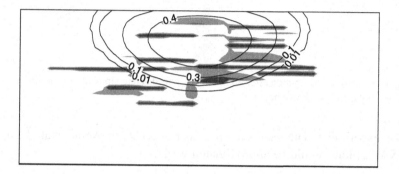

Figure 15. Comparison between numerical results with the resolved lenses (grey scales) and the upscaled parameters (contour lines).

In order to compare the results we calculated the first and seconds moments for the three different set-ups at three different times. Figures 16 and 17 show the saturation distributions after 20 and 40 minutes, respectively. The origin of the coordinate system is located at the midpoint of the top boundary, with the z-coordinate pointing downwards. The moments are given in Table 2. As the geometric-average and the upscaled configuration are obviously symmetric with respect to the z-axis the first moment in x-direction is not shown. The moments calculated for the geometric averaging case after 40 and 60 minutes are written in brackets, because at that time the DNAPL is already pooling at the bottom of the domain.

Table 2 shows that the results for the upscaled simulation are better than in the geometric averaging case. In the geometric averaging case the moments are significantly overestimated in the z-direction and underestimated in the x-direction. The moments for the upscaled case still underestimate all spatial moments of the discrete case. This is expected for the second moment in x-direction, where the underestimation is most pronounced, as the lenses transport DNAPL more efficiently to boundary regions than can be captured by the effective upscaled parameters.

6. Final Remarks

We have presented a quick and simple upscaling technique for DNAPL infiltration at the laboratory scale. We have applied the method to experimental data of a

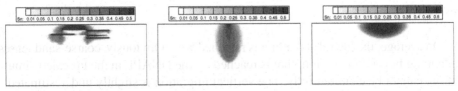

Figure 16. Saturation Distribution after 20 minutes – left: discrete simulation – middle: geometric averaging – right: upscaled parameters

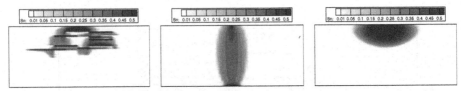

Figure 17. Saturation Distribution after 40 minutes – left: discrete simulation – middle: geometric averaging – right: upscaled parameters

sandbox experiment. The results are promising, since the overall spatial extent of the DNAPL plume could be approximated well.

Our approach consists of a percolation model to obtain the macroscopic P_c-S-relationship and of a single-phase flow-approach to determine the effective permeabilities as a function of mean saturation. Currently, we use a site percolation model, which should be replaced by an invasion percolation model in the near future. The single-phase flow-model for the upscaling of relative permeability is especially useful when the system is strongly anisotropic, and the renormalization approach would fail. In the current application, both the single-phase flow-model and the renormalization approach yield strong macroscopic residual saturations and anisotropic behaviour as shown earlier by e.g. (Pickup and Sorbie, 1996).

The presented upscaling approach is subject to the following underlying assumptions:

- Capillary equilibrium is assumed.
- The fluctuations of the flow velocities and the parameter functions are neglected in the dimensional analysis.
- We have not upscaled the form of the equation but determined effective parameters assuming that the form of the equation is conserved.

The approach is therefore restricted to capillary dominated systems. Also this upscaling method is only applicable to the specific scales used in here.

Especially the capillary equilibrium assumption needs further analysis. The method should also be compared to homogenization theory (cf. e.g. (Duijn et al., 2002)). Further examinations of the influence of different heterogeneities on multi-phase flow, e.g. pooling and the influence of lenses needs investigations

Table 2. First and second moments of the DNAPL body.

time	20 min	40 min	60 min
1. moment z-direction [m]:			
discrete	0.0650	0.1127	0.1582
geometric	0.1421	(0.2472)	(0.3388)
upscaled	0.0472	0.0724	0.0993
2. moment z-direction [m^2]:			
discrete	0.0016	0.0039	0.0059
geometric	0.0069	(0.0195)	(0.0220)
upscaled	0.0011	0.0023	0.0035
2. moment x-direction [m^2]:			
discrete	0.0181	0.0276	0.0381
geometric	0.0024	(0.0037)	(0.0131)
upscaled	0.0087	0.0160	0.0229

which should be accompanied by more laboratory experiments.

Up to now, only the main axis of a full $k_r - S$ tensor is implemented in the numerical code. An extension to include the full tensor is planned in the near future.

Acknowledgements

This work was supported by the Deutsche Forschungsgemeinschaft (DFG) in the framework of the project MUSKAT (He-2531/2-2) and the Emmy-Noether program (Ne 824/2-1, Ci 26/3-4). We are grateful to the team of the VEGAS facility for their cooperation, and to the two referees for their useful comments and suggestions.

References

H. Kobus, U. de Haar. *Perspektiven der Wasserforschung*, DFG, 1995.
J. Allan, J. Ewing, J.Braun, and R. Helmig. *Scale Effects in Multiphase Flow Modeling*, In First International Conference on Remediation of Chlorinated and Recalcitrant Compounds, Non-aqueous Phase Liquids, G. B. Wickramanayake and R. E. Hinchee Monterey / California, USA, 1998.
H. Kobus, J. Braun, and J. Allan. *Abschlussbericht 'Parameteridentifikation in Mehrphasensyste-men*. Scientif Report HG 274 WB 00/07, Institut für Wasserbau.
H. Jakobs, R. Helmig, C. T. Miller, H. Class, M. Hilpert, and C. E. Kees. *Modelling of DNAPL flow in saturated heterogeneous porous media*, Preprint 25, Sonderforschungsbereich 404, Mehrfeldprobleme in der Kontinuumsmechanik, Universität Stuttgart, 2003.

T. H. Illangasekare, J. L. Ramsay, K. H. Jensen, and M. B. Butts *Experimental study of movement and distribution of dense organic contaminants in heterogeneous aquifers*, Journal of Contaminant Hydrology, 20 (1-2): 1 – 25, 1995.

M. A. Christie, *Upscalig for Reservoir Simulation*, Journal of Petroleum Technology, 48 (11): 1004 – 1010, 1996.

G. E. Pickup, K. S. Sorbie. *The scaleup of Two-Phase Flow in Porous Media Using Phase Permeability Tensors*, SPE Journal, December 1996.

M. Dale, S. Ekrann, J. Mykkeltveit, and G. Virnovsky. *Effective Relative Permeabilities and Capillary Pressure for One–Dimensional Heterogeneous Media*. Transport in Porous Media, 26: 229–260, 1997.

Y. C. Chang and K. K. Mohanty. *Scale–up of two–phase flow in heterogeneous porous media*. Journal of Petroleum Science and Engineering, 18: 21 –34, 1997.

A. J. Desbarats. *Upscaling capillary pressure–saturation curves in heterogeneous porous media*. Water Resources Research, 31(2): 281 – 288, 1995.

T. C. J. Yeh, L. W. Gelhar, and A. L. Gutjahr. *Stochastic Analysis of Unsaturated Flow in Heterogeneous Soils 3 . Observations and Applications*. Water Resources Research, 21(4): 465 – 471, 1985.

A. Mantoglou and L. Q. Gelhar. *Stochastic Modelling of Large–Scale Transient Unsaturated Flow Systems*. Water Resources Research, 23(1): 37–46, 1987.

C-M. Chang et al. *Stochastic analysis of two-phase flow in porous media: I. Spectral/perturbation approach*, Transport in Porous Media, 19, 233-259, 1995.

I. Neuweiler, S. Attinger, W. Kinzelbach, and P. R. King. *Large Scale Mixing for Immiscible Displacement in Heterogeneous Porous Media*, Transport in Porous Media, 51: 287 – 314, 2003.

Y. Efendiev and L. J. Durlofsky. *A Generalized Convection-Diffusion Model for Subgrid Transport in Porous Media*, submitted to SIAM MMS, 2003.

M. Quintard and S. Whitaker. *Two–Phase Flow in Heterogenous Porous Media: The Method of Large–Scale Averaging*. Transport in Porous Media, 3: 357–413, 1988.

A. Ahmadi and M. Quintard. *Large-scale properties for two-phase flow in random porous media*, Journal of Hydrology, 183: 69 – 99, 1996.

A. E. Sáez, C. J. Otero, and I. Rusinek. *The effective homogenous behaviour of heterogenous porous media*. Transport in Porous Media, page 213–238, 1989.

A. Bourgeat and M. Panfilov. *Effective two–phase flow through highly heterogeneous porous media: Capillary nonequilibrium effects*, Computational Geosciences, 2, 191 – 215, 1998.

Y. S. Wu and K. Pruess. *A Multiple–Porosity Method for Simulation of Naturally Fractured Petroleum Reservoirs*. Society of Petroleum Engineers, SPE 15129, page 335–349, 1986.

K. Pruess. *On the Validity of a Fickian Diffusion Model for the Spreading of liquid Infiltration Plumes in Partially Saturated Heterogeneous Media*. Computational Methods in Water Resources, 10: 537 – 544, 1994.

K. Pruess. *A Fickian Diffusion Model for the Spreading of Liquid Plumes Infiltrating in Heterogeneous Media*. Transport in Porous Media, 24: 1–33, 1996.

A. T. Corey and C. H. Rathjens. *Effect of Stratification on Relative Permeability*, Journal of Petroleum Technology, 69 – 71, 1956.

E. H. Smith. *The Influence of Small–Scale Heterogeneity on Average Relative Permeability*, Reservoir Characterization II, 52 – 76, 1991, Academic Press, Inc.

S. Ekrann and M. Dale and K. Langaas and J. Mykkeltveit. *Capillary Limit Effective Two–Phase Properties for 3D Media*, Society of Petroleum Engineers, SPE 35493, 119 – 129, 1996.

Y. C. Yortsos, C. Satik, J. C. Bacri, and D. Salin. *Large-scale percolation theory of drainage*. Transport in Porous Media, 10: 171–195, 1993.

B. H. Kueper and B. Girgrah. *Effective large–scale parameters for two–phase flow in heterogeneous porous media*. In Transport and Reactive Processes in Aquifers, page 531 – 536. Dracos & Stauffer, 1994.

T. R. Green, J. E. Constanz, and D. L. Freyberg. *Upscaled soil–water retention using van Genuchten's function*. Journal of Hydrologic Engineering, 1(3), 1996.

G. E. Pickup, and K. D. Stephen. *An Assessment of steady-state scale-up for small-scale geological models*, Petroleum Geoscience, Vol. 6, 203 – 210, 2000.

C. Braun, R. Helmig, and S. Manthey. *Determination of constitutive relationships for two–phase flow processes in heterogeneous porous media with emphasis on the relative permeability–saturation–relationship*, Journal of Contaminant Hydrology, 67(1-2): 47-85, 2003.

C. M. Marle. *Multiphase Flow in Porous Media*, 1. edition, Institut Francais du Petrole, Paris, 1981.

R. Helmig. *Multiphase Flow and Transport Processes in the Subsurface*, Springer-Verlag, Heidelberg, 1997.

C.J. van Duijn, A. Mikelic, and I.S. Pop. *Effective Equations for Two-Phase Flow with Trapping on the Micro Scale*, SIAM J. Appl. Math., 62(5): 1531-1568, 2002.

D. Stauffer. *Introduction to Percolation Theory*, Taylor & Francis, London, 1985.

J. K. Williams. *Simple Renormalisation Schemes for Calculating Effective Properties of Heterogeneous Reservoirs*, in 1st European Conference on the Mathematics of Oil Recovery, Cambridge, UK, 1989.

P. R. King. *Upscaling Permeability: Error Analysis for Renormalisation*. Transport in Porous Media, 23: 337–354, 1996.

R. H. Brooks, and A. T. Corey. *Properties of porous media affecting fluid flow*, Journal of the Irrigation and Drainage Division Proceedings of the American Society of Engineers, 1966, volume 92, IR2, pages 61 - 88.

B. B. Dykaar and P. K. Kitanidis *Determination of the Effective Hydraulic Conductivity for Heterogeneous Porous Media Using a Numerical Spectral Approach. 1. Method*, Water Resour. Res., 28 (4), pages 1155-1166 , 1992.

L. J. Durlofsky. *Numerical Calculation of Equivalent Grid Block Permeability Tensors for Heterogenous Porous Media*, Water Resources Research, 27(5): 699 – 708, 1991.

C. Braun *Ein Upscaling Verfahren für Mehrphasenströmungen in porösen Medien*, Forschungsbericht 103, Ph.D. Thesis, Mitteilungen des Instituts für Wasserbau, Universität Stuttgart, Stuttgart, 2000.